MARINE LIFE OF SOUTHERN CALIFORNIA

Emphasizing Marine Life of Los Angeles-Orange Counties

Second Edition

by
DONALD J. REISH, PhD

Emeritus Professor of Biology
California State University, Long Beach
Long Beach, California

Illustrated by
Lhisa J. Reish
Janice Findley Fisher

KENDALL/HUNT PUBLISHING COMPANY
4050 Westmark Drive   Dubuque, Iowa 52002

Front Cover:

Point Fermin, California, looking south. Lin Craft, photographer.

Back Cover:

Top photo: The green sea anemone, *Anthopleura xanthogrammica* and the purple sea urchin, *Strongylocentrotus purpuratus.* Suzanne Miller, photographer.

Middle photo: The giant kelp, *Macrocystis pyrifera.* Lin Craft, photographer.

Bottom photo: *Allopora Californica.* Lin Craft, photographer.

Copyright © 1972, 1995 by Donald J. Reish

Library of Congress Number: 95-78048

ISBN 0-7872-1045-5

Printed in the United States of America
10 9 8 7 6 5 4 3 2

DEDICATION

Looking back over my career which spans nearly 50 years of graduate and professional teaching and research in marine biology, many people and events have played a role in my life. I dedicated the first edition of "Marine Life of Southern California" to the professors who especially played a significant role in my training. I shall always be appreciative of their assistance; these professors were the late Dr. Olga Hartman, the late Dr. Ivan Pratt, and Dr. John L. Mohr. There is another group of people who also played a significant role in the professional career; these were the 57 graduate student who I had the opportunity to guide through their Masters' Degree thesis research and the many students who enrolled in my classes at California State University, Long Beach. The association with these students over the years make teaching perhaps the most rewarding of all careers.

TABLE OF CONTENTS

PREFACE TO THE FIRST EDITION

The author believes that the name of the organism is essential to further understanding and appreciation of the science of marine biology. This statement applies equally well to the layman or to the professional biologist. Many books and pamphlets have been written on the marine organisms of the Pacific Coast since the classic by Johnson and Snook in 1927. In addition there have been many extensive studies dealing with one or more groups of organisms; these are frequently highly technical in nature and are of limited use to the layman. Most books have covered a wide geographical area. The author believed that a need existed for a book dealing with the marine plants and animals of a limited geographical area in greater detail. With these objectives in mine, the author chose to describe some of the plants and animals known to occur in Los Angeles and Orange Counties, California. It is the hope of the author that it will be possible for the layman to be able to identify many of the plants and animals he might encounter in these counties with the aid of the figures, descriptions and discussions in this book. While the specific localities mentioned are areas where the author or his students have actually seen the particular organism, it is quite possible that they will be seen at other localities in Los Angeles and Orange Counties. While this book deals specifically with these two counties, it will be of use to the north and south. However, as one gets farther from this area in southern California (200-300 miles), the value of this reference is diminished.

The author wishes to call attention of the reader to the section on "How to Use This Book" and "Conservation of Marine Life" in the Introduction. During certain times of the year or in certain years, marine organism may appear and later disappear. The triggering mechanism for the sudden appearance of a particular species is generally unknown; it is undoubtedly due to the interplay of chemical, physical and biological factors. It is possible then for what is normally a rare organism, extremely common occurrences may arise. Because of this, various organisms will not be discussed herein due to the usual rarity of the plant or animal.

Illustrations are essential to a book of this nature. The author wishes to express his thank to Mrs. Janice Findley Fisher for making these illustrations. They were made from either living or preserved material. Additional thanks are due to many of my students in marine biology who have assisted me in collection of material and encouraged me in many ways.

Many people has assisted the author in the identification of marine organisms over a period of years; however, the identifications used herein are the responsibility of the author and not necessarily those of the specialists. The appreciation of their help is deeply acknowledged; they are, Mr. Ira Cornwall (barnacles), Dr. J. Laurens Barnard (amphipods), Drs. John D. and Dorothy Soule (ectoprocts), Dr. Donald P. Abbott (tunicates), Dr. E. D. Lane (fishes), Dr. Charles T. Collins and Dr. Stuart L. Warter (birds), the late

Dr. E. Yale Dawson and Dr. Thomas Widdowson (algae) and Dr. Philip C. Baker (marine grasses).

This book is largely an outgrowth of the "Marine Life of Alamitos Bay" written by the author in 1968. The format utilized in that book has been followed herein. The author wishes to thank the many people for their kind words of encouragement and support for that book.

Donald J. Reish
January 1972

PREFACE TO THE SECOND EDITION

While I was writing the first edition of this book, I decided to publish the book personally. The primary reason for this decision is that I wanted the material in the book presented in a particular way; that is, I wanted the figure of the organism next to the description and discussion of that plant or animal. I believe that this makes the book more useful to the reader. The reader doesn't have to flip pages back and forth while trying to determine which species he/she is holding. Judging from the response that I have received over the years from people, I believe this proved to be the best way to present such material. I have followed this same procedure in the second edition. In 1982 I turned over the publication and marketing of this book to Kendall/Hunt Publishing Co. They have followed this same method of presentation of material in the second edition.

The second edition has been enlarged over the first edition in many ways. The introduction has been expanded to included topics which are of greater interest today than two decades ago. A key to the major groups of organisms has been included which will facilitate the reader in determining which of the many groups of organisms his/her particular specimen belongs. Over 100 more organisms have been added especially the larger species such as decapods, fish and birds. Additional information about the biology of the animal, such as food habits, is included, whenever it is known.

Many of the illustrations in the introduction are either new or have been redrawn. Some of the existing drawings of organisms were redrawn in addition to those newly added species. It is a pleasure to thank my daughter, Lhisa J. Reish, for making these illustrations.

I wish to express my thanks to the management of Kendall/Hunt for their encouragement to undertake a second edition of this book and to honor my method of presentation of the material.

Donald J. Reish
December 1994

INTRODUCTION

The marine environment of the Los Angeles—Orange Counties, California is situated in the middle of the region referred to as the Southern California Bight (Fig. 1). The Bight includes an area of over 3000 square miles which extends from Point Conception south into Baja California. The continental shelf is located some 180 miles offshore, however there are many basins between the mainland and the shelf the deepest of which is over 9000 feet. Sandy beaches dominate the coastline in Los Angeles—Orange Counties with rocky shores limited to Malibu, Palos Verdes Peninsula (Fig. 4), Corona del Mar (Fig. 9), Laguna Beach (Fig. 9), and rocky jetties constructed by mankind.

The population of these two counties totals about 15 million people which is expected to double in the next 4 or 5 decades. The mild Mediterranean climate experienced in Southern California is one of the primary attractions of this region. The beaches and offshore waters are important year round recreational areas for many of the inhabitants. The coastline of these counties has been and is being altered by human activity. The only nearly unaltered areas of coast lie with the military bases of Point Mugu and the Seal Beach Naval Weapons Depot (Fig. 7) and the Bosa Chica Slough (Fig. 7) and portions of upper Newport Bay (Fig. 8). Some areas, such as Marina del Rey (Fig. 3), Alamitos Bay (Fig. 6) and lower Newport Bay (Fig. 8) have been altered for housing and recreation. Los Angeles—Long Beach Harbors (Fig. 5) were developed from former wetlands for industrial and shipping purposes. Ballona Creek, Los Angeles River (Fig. 5), San Gabriel River (Fig. 5) and Santa Ana River have been dammed at the head waters, channelized and altered for flood control. The areas located off El Segundo, White's Point (Fig. 4) and near the mouth of Santa Ana River have been altered by the discharge from the large domestic sewage treatment plants. Four offshore islands were constructed off Long Beach for oil drilling and offshore oil platforms are located off Huntington Beach. However, in spite of the alterations to the marine environment by mankind, the coastal region of Los Angeles—Orange Counties still contains a diversity of marine organisms, probably between 1000-2000 or more different species of macroscopic plants and animals in the intertidal to shallow offshore waters.

PHYSICAL AND CHEMICAL CHARACTERISTICS OF THE MARINE WATERS OF THE SOUTHERN CALIFORNIA BIGHT

TIDES

There are two high tides and two low tides each day in Southern California with approximately 6 hours between each high and low tide. The tides are predicted and published each year by the U. S. Hydrographic Office and local tide books can be obtained from marine hardware or sporting good stores. Many newspapers published daily tidal information. The movement of the tides is controlled primarily by the gravitational attraction of the moon and to a lesser extent by the sun. The greatest tidal fluctuations occur when the moon is in

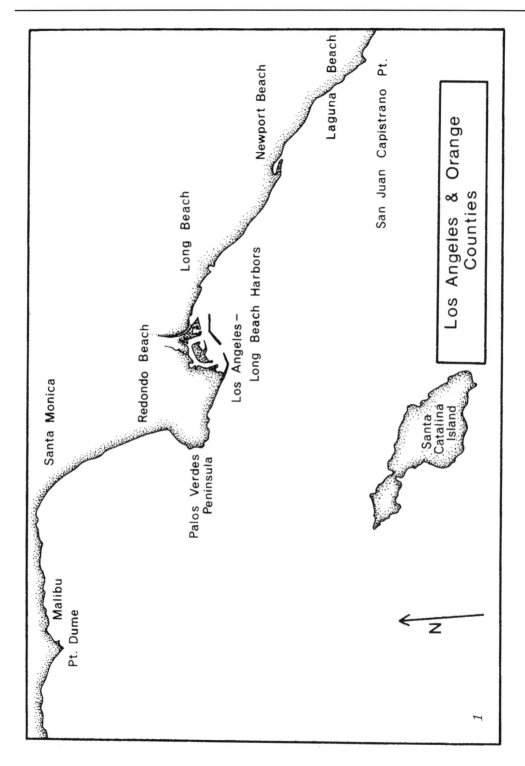

Los Angeles & Orange Counties

1

either full or new moon phases (see below). These tides are commonly referred to as spring tides which has no relationship to the season of the year. Tidal conditions are said to be flat when the moon is in either first or last quarter (see below). These tides are commonly called neap tides.

Full moon		Full moon plus 1 day	
0340 hours	0.6 ft	0433 hours	0.5 ft.
0946	6.1	1035	5.4
1619	-1.4	1653	-1.1
2237	5.3	2318	5.4

First quarter	
0256 hours	4.6 ft
1109	1.2
1751	2.8
2025	2.5

In the example given above the maximum tidal fluctuation on the day of full moon was 7.5 feet as compared to 3.4 feet during the first quarter. In general one particular tide occurs 24 hours and 50 minutes later each day although this time difference between tides on successive days does not always hold true as noted in the example given above for the full moon tides. The greatest diversity of marine life can be seen during extremely low tides or minus tides. It is best to visit such areas an hour or two before these very low tides to afford a longer period for observation of marine life. Minus tides occur during the daylight hours from about September to March, and during the night from March to September. Certain species of marine plants and animals can be used as indicators for particular tidal zone; such information is useful in judging the extent of a particular low tide. See the discussion of zonation below for additional details.

CURRENTS AND WAVES

Currents are defined as major horizontal movements of water which are continuous and in a definable path. The California current (Fig. 2) is the principle one of the Southern California coast. It is a cold water mass which comes across the North Pacific Ocean and turns southward off the coast of British Columbia. The principle component of the California current follows the outer edge of the continental borderland. It flows southward to near the equator where is turns westward. The California current is a few hundred miles wide and flows as much as one-half miles per hour. The extent of this current can be measured since the surface water temperature is a few degrees cooler than the surrounding water. In the central portion of the Southern California Bight a smaller counter current, the

Southern California countercurrent, flows northward in a counterclockwise direction until it is substantially blocked by the northern Channel Islands. A smaller component of this countercurrent continues north of Point Conception and is termed the Davidson current. An offshore wind-driven current may occur during periods of "Santa Ana" winds. This offshore wind drives the surface water out to sea forcing colder, deeper water to the surface near shore. This upward movement of colder waters is referred to as upwelling and it plays an important role in bringing nutrients from the deeper water to the surface.

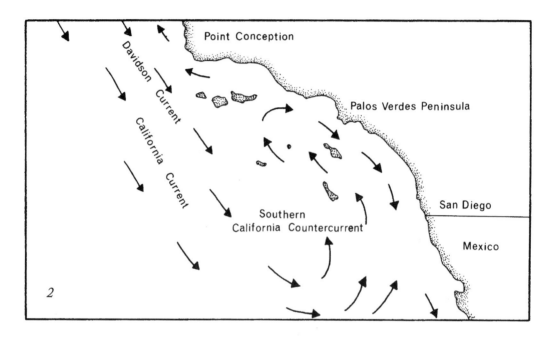

Waves are defined as disturbances which move either on or in the water mass. They may be caused by a local storm or a disturbance thousands of miles away. As every surfer knows, some local beaches have a better surf, such as Huntington Beach, than others; this largely depends on the particular location of the beach with reference to such factors as islands, breakwaters, kelp beds, local geography, etc.

EL NIÑO

The term El Niño, or "The Child", was apparently coined by the Peruvian fishermen because of its yearly occurrence off the coast of Peru around Christmas time. Typically the Peruvian current extends northward into the Northern Hemisphere and the Equatorial Countercurrent is displaced to the south. These waters are warmer and lower in salinity and converge with the Peruvian current and move south. The peak of this yearly event occurs in February and March and usually does not affect the oceanographic conditions in Southern California. Many physical changes occur which causes modifications of the current pattern and oceanic environment. A decrease in the upper westerly winds in the south western Pacific Ocean leads to a difference in sea level between the two sides of the Pacific Ocean at the equator. In some years the California current is weakened and the Equatorial Countercurrent is diverted northward. This brings warmer and lowered salinity waters into the Northern Hemisphere and Southern California. The warmer waters off Southern California causes changes in the biota especially in the planktonic forms. While El Niño is a yearly event in Peru, it occurs about every 4 to 10 years and is believed to be caused by the differences in sea level atmosphere between Easter Island and Darwin, Australia. In some years a contrasting phenomenon occurs in which the water is colder than normal. This condition is termed La Niña ("The Girl") and is caused by stronger currents from the north and weaker ones from the south. This occurrence is noted about every four years.

WATER TEMPERATURE

The water temperature of offshore waters varies from a January-February of 50-52° F. to an August-September high around 75° F. Water temperatures in bays and harbors is generally a few degrees warmer. This temperature fluctuates from year to year and is influenced by the water temperatures of the major ocean currents, degree of upwelling, and potential effects of El Niño or La Niña. The water temperature may be influenced locally by wastes discharges from the large sanitation districts or heated cooling waters from electrical generating plants. Fluctuations in the water temperature influence the species composition of planktonic and fish populations but probably has little or no effect on subtidal benthic populations.

CHEMICAL CHARACTERISTICS

The salinity of the water is that of normal sea water (about 34-36°/$_{oo}$) throughout much of the Southern California Bight except following winter rains. Since rains are infrequent in Southern California and the recovery to normal salinity is rapid (a day or two after cessation of rainfall), the salinity is that of normal sea water even within the bays and harbors.

The dissolved oxygen content of the water is generally well over 5.0 mg/l, the minimum amount considered necessary to sustain fish and aquatic life. Values below 6.0 mg/l are measured in areas such as bays and harbors where water circulation may be limited. These lower figures may be the result of oxygen depleting wastes from man's activities (pollution) coupled with reduced water circulation; they may be caused by some natural phenomenon such as a die-off from a red-tide (see below). Whatever the cause, long term exposure to reduced concentrations of dissolved oxygen below 5.0 mg/l may lead to a reduction in species diversity.

The dissolved nutrients, namely phosphates, nitrates, nitrites, and silicates are important chemical compounds in the water especially in the growth of diatoms, the primary food source in the marine food chain. These chemicals are necessary for completion of the photosynthetic process. The concentration of these nutrients will vary according to water depth, distance from shore, season (rainfall), currents, winds, location of rivers and to discharges from man's activities. These values, away from the sources of pollution, average 0.012 to 0.165 micro-gram atoms per liter (g-at./l) for phosphates and 0.05 to 0.94 g-at./l for silicates. The values for nitrates and nitrites vary according to such factors as water depth. Surface figures for outer Los Angeles Harbor averaged 9.1 g-at./l for nitrates and 8.9 g-at./l for nitrites. Nitrates were 2.4 g-at./l for nearby Alamitos Bay.

THE MARINE ENVIRONMENT OF LOS ANGELES-ORANGE COUNTIES

The coastline of Los Angeles-Orange Counties is a series of rocky shores and sandy beaches interspersed with bays and harbors (Fig. 1). The rocky shores are bordered by cliffs and the sandy beach, bays and harbors by low-lying lands.

ROCKY SHORES

Rocky shores are present in the vicinity of Pt. Dume with accessible beaches at Pt. Dume, Zuma Beach and Malibu Beach (Fig. l). The Palos Verdes Peninsula has the most extensive rocky shore areas of Los Angeles County (Figs. 1, 4). The more accessible rocky beaches include Flatrock Pt., Pt. Vincente, White's Pt. and Pt. Fermin (Fig. 4).

Little Corona (Fig. 8) is a convenient rocky shore to visit in Orange County. The Laguna Beach area is a series of small rocky shores and coves of sandy beaches. The rocky shores of these two counties are marine life preserves. Fishing is permitted but the collection of other marine organisms is forbidden.

SANDY BEACHES

Los Angeles-Orange Counties are blessed with many miles of sandy beaches which are ideal as a year round playground. The two major sandy beach areas include the region of Santa Monica Bay which extends from Malibu to the Palos Verdes Peninsula (Figs. 1, 3). This stretch of sandy beach is interrupted by boat harbors at Marina del Rey, Ballona Creek, King's Harbor and an occasional rock jetty or pier. The second major sandy beach extends from Long Beach to Newport Beach (Fig. 1). This beach is interrupted by Alamitos Bay (Fig. 6), Anaheim Bay (Fig. 7) and the mouth of Santa Ana River. Smaller sandy beaches are found between the rocky outcrops from Corona del Mar to Dana Pt. (Fig. 9).

SHALLOW OFFSHORE WATERS

The intertidal rocky shores extend into subtidal waters and are an important habitat for the giant kelp and the associated algae, fish and invertebrates. The particle size of the sediments become smaller in size as the water depth increases. Sediments consist of fine sands, silts and clays and provide habitat for hundreds of different species of clams, snails, crustaceans, worms and fish some of which are illustrated in this book.

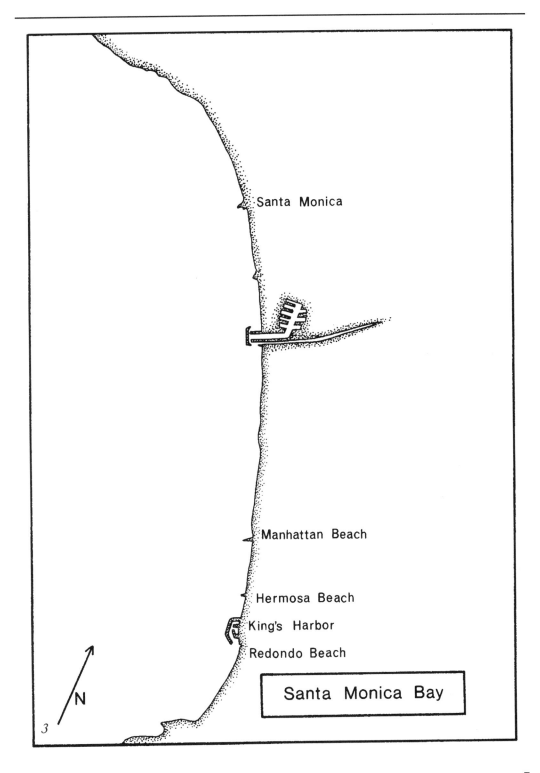

Santa Monica

Manhattan Beach

Hermosa Beach

King's Harbor

Redondo Beach

3

N

Santa Monica Bay

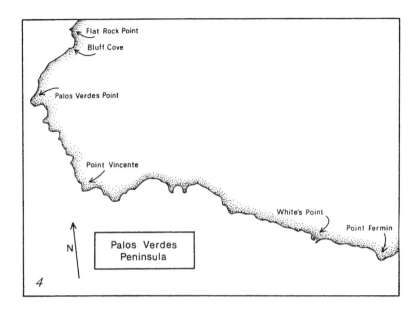

Flat Rock Point

Bluff Cove

Palos Verdes Point

Point Vincente

White's Point

Point Fermin

N

Palos Verdes
Peninsula

BAYS AND HARBORS

An estuary is defined as a body of water which contains water of varying degrees of salinity. Estuaries are generally associated geographically with the mouths of rivers. Southern California does not have a typical estuary characteristic of most coastal areas. This is primarily because of the small amount of fresh water run-off from rivers because of the infrequent winter rains and because so many of the small mountain streams have been dammed in Southern California for flood control. The salinity of most of the bays decreases soon after the rains fall but recovery to normal salinity is generally within a day or two following the cessation of rainfall. Therefore, all the bays and harbors in Southern California contain normal concentrations of sea salts most of the time.

The distinction between a bay and harbor is not well defined. A harbor is generally referred to as a body of water which has been greatly altered by man and is used primarily for shipping. Los Angeles-Long Beach Harbors, which are really one harbor oceanographically, is a highly industrialized area which is used for commercial shipping, a military naval base and for recreation (Fig. 5). A bay is an area with normal salinity which may or may not be modified by man. Alamitos Bay (Fig. 6) and Newport Bay (Fig. 8) are similar in that they are primarily recreational bodies of water used for boating, swimming and bathing. Both of these bays were largely marshlands within the past 100 years much like Anaheim Bay and Bolsa Chica Slough (Fig. 7) are today. Huntington Harbour (Fig. 7) was developed in the 1960-1970's as a residential boating complex from marshlands. Marina del Rey

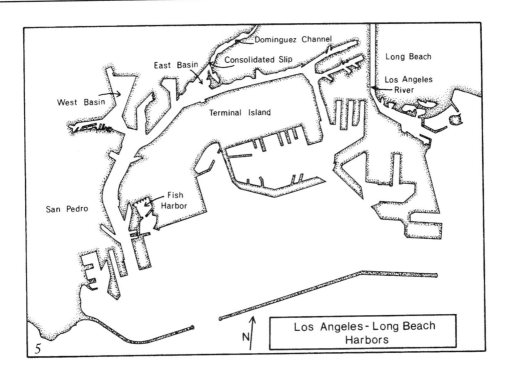

(Fig. 3) was dredged and developed from ancient sand dunes into a large multi-storied housing, boating and commercial area. King's Harbor (Fig. 3) and Dana Pt. Marina (Fig. 9) were constructed by extending rock jetties out from the shore.

The marshlands of Anaheim Bay (Fig. 7). Bolsa Chica Slough (Fig. 7) and portions of upper Newport Bay (Fig. 8) are the last remaining essentially undisturbed protected embayments in Los Angeles-Orange Counties. These areas are important as fish nurseries and the primary feeding and resting areas for resident and migratory birds in these two counties.

MARINE BIOLOGY OF
LOS ANGELES—ORANGE COUNTIES

ZONATION

Zonation refers to the occurrence of a specific plant or animal at a particular water level. The tidal variation in Southern California is approximately 10 feet from the highest high tide to the lowest low tide. With the rise and fall of the tides generally two times each day, intertidal organisms are exposed to the air for varying periods of time depending upon the water level they inhabit and the extent of the particular tide. Since each species has its own characteristic requirements for food, light, oxygen, nutrients, substrate, etc., they live in that

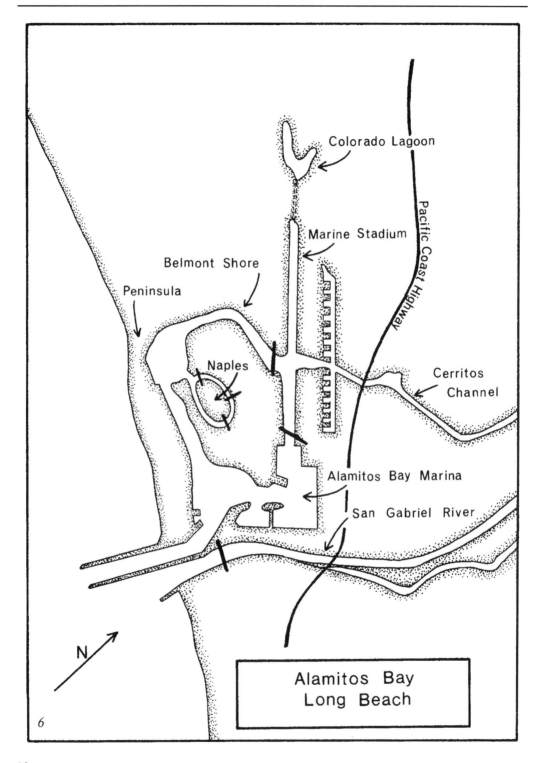

Colorado Lagoon

Marine Stadium

Belmont Shore

Peninsula

Cerritos Channel

Naples

Alamitos Bay Marina

San Gabriel River

Pacific Coast Highway

N

6

Alamitos Bay
Long Beach

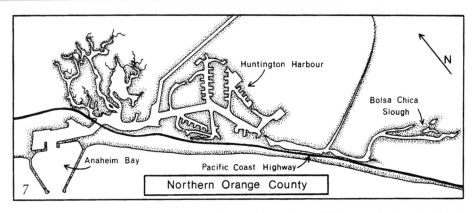

zone where these requirements are optimum for them. This results in the horizontal groupings of organisms or zonation (Figs. 10-13). Certain species, because of their size, numbers of individuals present, their well-defined limits within the intertidal environment, or their ease in recognition, are indicators of a particular tidal zone. Thus, for example, the California mussel, *Mytilus californianus* (Figs. 10, 12, 201, 202), and the bay mussel, *Mytilus edulis* (Figs. 11, 203, 204), are indicative of the mid-tide horizon of rocky shores and bay environments, respectively.

While zonation is observed easiest at rocky shores and on pilings, it occurs at sandy beaches, back bays (Fig. 13), and in subtidal waters. Zonation in sandy beaches is difficult to observe because of the animals live within the sand. Zonation in back bay areas (Fig. 13) is recognized by many different species of plants and by the California horn snail, *Cerithidea californica* (Fig. 137). Zonation in subtidal waters is govern by physical factors such as substrate, water depth, and light.

Zonation in intertidal waters is determined by the tidal cycle which is conveniently divided into the splash zone, high tide zone, mid-tide zone, low tide zone, minus tide zone and the subtidal zone. The need for water and the associated biological, chemical and physical environmental factors increases as one proceeds down the tidal horizon. The organisms living in the splash zone are rarely covered by water. Much of their needs for water are met by water splashing over them. The height of the splash zone depends upon local physical conditions. If the splash zone is situated along an open, rugged coast, such as the Palos Verdes Peninsula, then the splash zone may extend many feet above the high water line because of the heavy surf and often associated winds. If, on the other hand, the splash zone is located within the quiet waters of a bay, then this zone is quite narrow because of the absence of a surf. The width of the other zones are not as variable since these organisms require daily submersion in the water for varying lengths of time.

Rocky Shores. A diagrammatic representation of intertidal zonation of a rocky shore is given in Figure 10. Rocky shores in Southern California characteristically have 5 distinct zones. It should be emphasized that zonation is not exactly the same on every rock face at a particular shore. Variation can occur depending upon where the rock is located with respect to the surf; thus, the rock surface facing the open sea can quite possible have a different

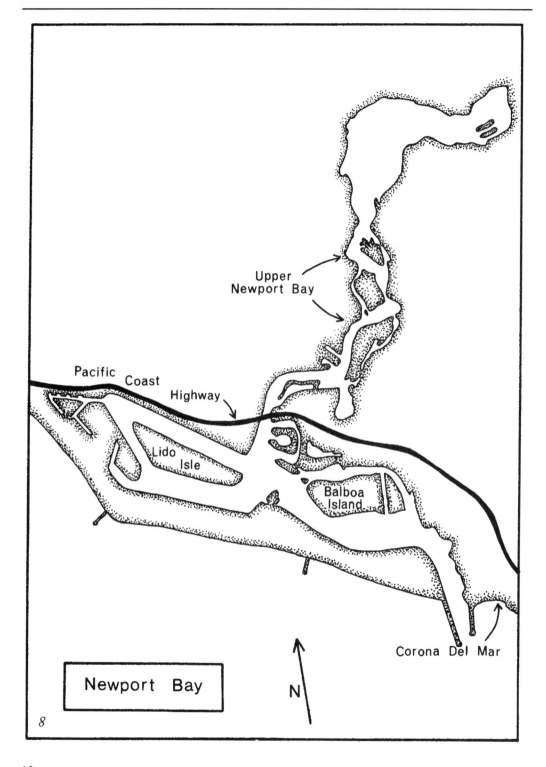

Upper
Newport Bay

Pacific Coast
Highway

Lido Isle

Balboa
Island

Corona Del Mar

Newport Bay

N

8

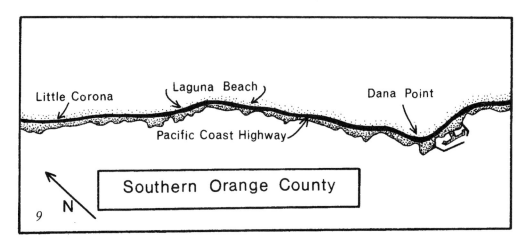

Little Corona

Laguna Beach

Dana Point

Pacific Coast Highway

Southern Orange County

N

assemblage of organisms than the surface facing the shore. Heavy seas following storms can denude a rock of its organisms which will take one or two years to be reestablished. The description of intertidal zonation given below presents the most frequently encountered type in Southern California and is intended to serve as a guide to an understanding of the biology of zonation.

The splash and high tide zones are characterized by barnacles, limpets and snails. Macroscopic algae does not occur in these zones unless there is fresh water drainage from cliffs. The barnacles, *Balanus glandula* (Figs. 10, 285) and *Chthamalus fissus* (Figs. 10, 12, 289), may be found scattered about anywhere on the rock. The gray colored *Chthamalus fissus* is more numerous than the larger, white colored *Balanus glandula*. These barnacles feed only when covered with water so they respond rapidly when splashed upon. The central part of the barnacle closes tightly when exposed to air which protects the animal from desiccation. The limpets *Collisella digitalis* (Figs. 10, 105, 106) and *Collisella limitula* (Figs. 10, 109, 110) are found generally in rock crevices along with the gray littorine, *Littorina planaxis* (Figs. 10, 130, 131). These animals generally move about during the night, sometimes down into a lower tidal zone, where they feed upon microscopic algae attached to the surface of the rocks. Frequently water or dampness is present within the crevice which minimizes desiccation. However, the limpets can fasten themselves securely to the rock surface and the littorine snail has an operculum to cover the soft parts within the shell; these features give them additional protection against desiccation. Two species of crustaceans are frequent visitors to the splash zone, the rock louse *Ligia occidentalis* (Fig. 298) and the striped shore crab *Pachygrapsus crassipes* (Fig. 345). *Ligia occidentalis* scurries about rapidly across rocks into crevices or protected places. *Pachygrapsus crassipes* wanders about different tidal zones and hides in rock crevices or under rocks.

The characteristic animal of the mid-tide zone is the California mussel *Mytilus californianus* (Figs. 10, 12, 201, 202). The width of this zone and the thickness of the clumps of mussels depend upon the slope of the rocks and the position of the rock with reference

INTERTIDAL ZONATION

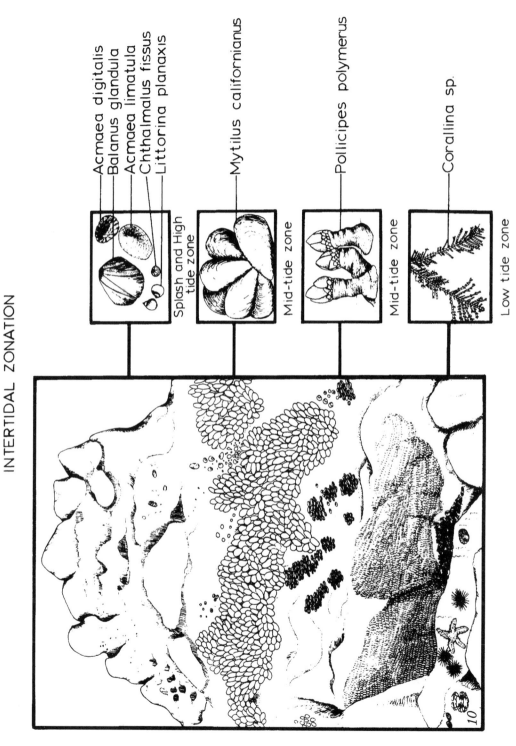

Acmaea digitalis
Balanus glandula
Acmaea limatula
Chthalmalus fissus
Littorina planaxis

Splash and High tide zone

Mytilus californianus

Mid-tide zone

Pollicipes polymerus

Mid-tide zone

Corallina sp.

Low tide zone

10

to the surf. Larger and thicker beds of mussels are found on more or less large, flat rocks which receive the full force of the oncoming waves. Under these conditions in Southern California, the bed of mussels may be nearly a foot in thickness and the individual mussel measures 6 inches in length. While the California mussel is the dominant organism in this zone, many species of plants and animals are associated with them. Clumps of the goose neck barnacle, *Pollicipes polymerus* (Figs. 10, 12, 280), may be found with a bed or beneath it (Fig. 10). Various species of acorn barnacles, limpets, sponges, polychaetes, ectoprocts and algae attach to the shells of the mussel. Representatives from nearly every invertebrate phylum are found attached or crawling within the mussel bed. Sponges, hydroids, sea anemones, flatworms, nemerteans, polychaetes, snails, limpets, pelecypods, copepods, isopods, amphipods, decapods, ectoprocts and an occasional starfish, sea urchin or brittle star are animals which can be seen within a mussel bed. Frequently green, brown and red algae are present. The California mussel affords protection to these animals from the full force of the waves as well as well as protection from enemies.

The low tide zone and minus tide zone are not as well defined. The most readily identified animal of these zones is the purple sea urchin *Strongylocentrotus purpuratus* (Fig. 389). It is often present in large numbers especially along horizontal rock shelves or in tide pools; if the rock is a soft sedimentary type, then the purple sea urchin will form burrows. The ochre starfish, *Pisaster ochraceus* (Fig. 384) is found in the low tide zone, but it often moves into the mid-tide zone to feed upon *Mytilus californianus*. Many species of brown and red algae are present, but the most characteristic one is a species of the articulate coralline algae *Corallina* sp. (Figs 10, 556). Two species of animals often build extensive tube masses along the sides of the rocks in these zones; these are the polychaete *Phragmatopoma californica* (Figs. 72, 73) and the scaly worm snail *Serpulorbis squamigerus* (Fig. 136).

Pilings. Zonation of marine organism can be readily seen on a marine piling since the structure is of uniform dimension and extends vertically through the different tidal zones. The organisms which attach to a piling will vary according to its location. A piling in offshore waters (Fig. 12) will have a different assemblage of organisms than one within a bay or harbor (Fig. 11). The species and thickness will vary on a piling depending upon amount of wave action and water quality; if the surf is heavy, the bed will be thick and the species diversity will be great. If the water quality is poor as a result, for example, pollution, then the number and diversity of organisms on the piling will be reduced.

The principle organisms attached to a piling have been diagrammatically represented for one in a bay (Fig. 11) and for one offshore (Fig. 12).

Bay pilings. For a piling located in a bay the splash zone is narrow because of the limited amount of wave action; the small acorn barnacle, *Chthamalus fissus* (Fig. 289), is the highest occurring organism found on these pilings. A second species of barnacle, *Balanus amphitrite* (Figs. 282, 283), appears at the lower limits of the splash zone and extends downwards into the high tide zone. A third species of acorn barnacle, *Balanus crenatus* (Fig. 284), appears in the high tide zone. The mid-tide zone is characterized by the appearance of the bay mussel,

INTERTIDAL ZONATION

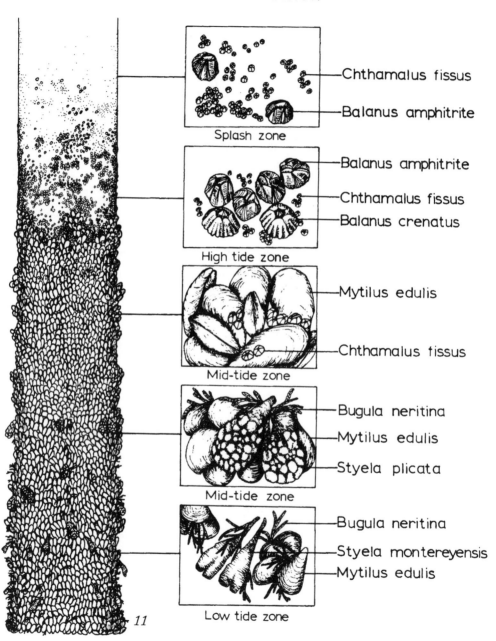

Chthamalus fissus
Balanus amphitrite
Splash zone

Balanus amphitrite
Chthamalus fissus
Balanus crenatus
High tide zone

Mytilus edulis
Chthamalus fissus
Mid-tide zone

Bugula neritina
Mytilus edulis
Styela plicata
Mid-tide zone

Bugula neritina
Styela montereyensis
Mytilus edulis
Low tide zone

11

Mytilus edulis (Figs. 203, 204), which is the dominant organism of a bay piling. Extensive clumps of this mussel attach to the piling and extend into subtidal waters. Acorn barnacles are found attached to the shells of the mussels but not in as large of numbers as found in the high tide zone. Several animals appear in the lower limits of the mid-tide zone such as the tunicate *Styela plicata* (Fig. 414), and the ectoproct *Bugula neritina* (363, 364). In the lower tide zone a second species of tunicate, *Styela montereyensis* (Fig. 413), appears. In addition to these conspicuous species many small worms, crustaceans, and other organism may be found among the clumps of mussels. The numbers of specimens and species of these smaller animals present within the clumps of mussels will be greater during the summer and fewer during the winter months.

Offshore pilings. The biological features of an offshore piling are diagrammed in Figure 12 from the low tide mark to subtidal depths. Different species of barnacles, limpets and algae are found above this tidal level. The amount of growth and diversity of biota on an offshore piling is much greater than present is one in a bay; growth of organisms especially in the mussel zone may be greater than one foot in thickness and contain over 100 species of algae and invertebrates. The California mussel, *Mytilus californianus* (Fig. 201, 202), is the dominant organism with specimens 8 to 10 inches in length. The ochre starfish, *Pisaster ochraceus* (Figs. 12, 384) are scattered up and down the piling where they feed upon the mussel. Different species of anthozoan cnidarians are attached to the piling or to the mussel; they include such species as the aggregating sea anemone *Anthopleura elegantissima* (Figs. 12, 35), the strawberry anemone *Corynactis californica* (Figs. 12, 38) and the white anemone *Metridium senile* (Fig. 12). Nearly every invertebrate phyla are found either attached to mussels or other organisms or are nestled within the protection of the mussels.

Sandy Beaches. Zonation occurs on open coast sandy beach but this is more difficult to see because of the fewer number of organism involved, and because of the constantly moving sands, the organism are generally covered. There are many microscopic animals (many smaller than an individual grain of sand) found in the spaces between damp to wet sand grains; but since they require special collecting techniques to study, they are not described herein. The upper reaches of the previous high tides are marked by the accumulation of dead algae and debris. Beach hoppers, such as *Orchestia traskiana* (Fig. 311) and *Orchestoidea californiana* (Fig. 312), are present in the sand of the high tide horizon especially wherever large quantities of seaweed have been washed ashore. During the day they will burrow into the sand, but they are more easily seen during the night with the aid of a flashlight; they crawl to the surface of the sand and hop about in search of food. The sand crab, *Emerita analoga* (Fig. 328), is not associated with any particular tidal zone but rather it moves up and down with the waves. In the mid-tide horizon such animals as the purple olive shell, *Olivella biplicata* (Fig. 181), the pismo clam, *Tivela stultorum* (Figs. 235, 236), the bean clam, *Donax gouldi* (Figs. 252, 253) and the polychaete *Nephtys californiensis* (Fig. 54), can be found just beneath the surface of the sand. In the low tidal zone many additional species of polychaetes, snails, clams, echinoderms and other groups may occur depending upon the degree of wave action and the size of the sand grains. A greater variety

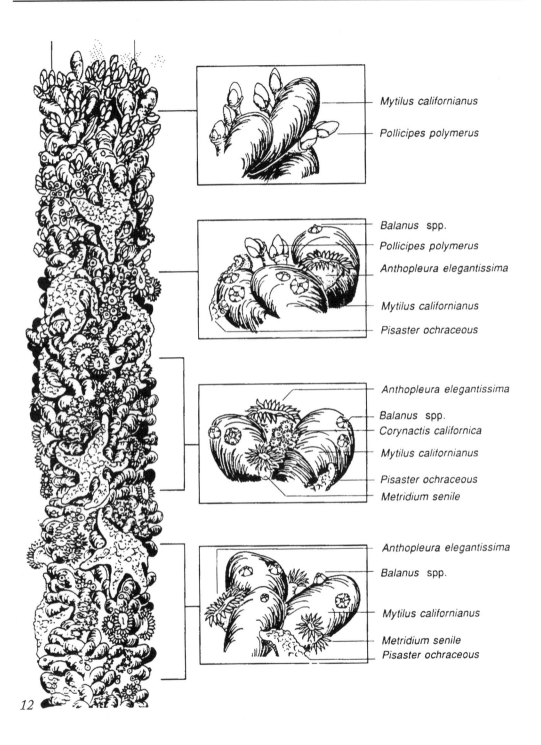

Mytilus californianus

Pollicipes polymerus

Balanus spp.
Pollicipes polymerus
Anthopleura elegantissima

Mytilus californianus

Pisaster ochraceous

Anthopleura elegantissima

Balanus spp.
Corynactis californica

Mytilus californianus

Pisaster ochraceous

Metridium senile

Anthopleura elegantissima

Balanus spp.

Mytilus californianus

Metridium senile
Pisaster ochraceous

12

of species will be present if the size of the sand grain is fine. Macroscopic algae are absent from a sandy beach because of the lack of a solid surface for attachment.

Back Bays. The sediment of back bay areas, for example Anaheim Bay and Newport Bay, is composed of silts and clays which do not move up and down with the tides such as occurs at sandy beaches. This permits the growth of grasses and flowering plants in the intertidal zone. The dominant plants of the back bays are represented diagrammatically by tidal zone in Figure 13. The zones are not as clearly deliminated as noted on pilings. The upper tidal zone above the plus 5 foot level is characterized by salt grass, *Distichlis spicata* (Fig. 576) and shore grass, *Monanthochloe littoralis* (Fig. 577). Other plants present in this zone include the sea-blite, *Suaeda californica* (Fig. 580), alkali heath, *Frankenia grandifolia* (Fig. 582) and marsh rosemary, *Limonium californicum* (Fig. 581). The saltwort, *Batis maritima* (Fig. 575) extends its runners both into the upper and middle tidal zones. Animals present in this zone include the yellow shore crab, *Hemigrapsus oregonensis* (Fig. 343) and the fiddler crab, *Uca crenulata* (Fig. 351) both of which dig burrows into the sediment. The ribbed horse mussel, *Geukensia demissus* (Fig. 200) can be found attached with their byssal threads to rocks or to the sides of the burrows made by the crabs.

Pickleweed, *Salicornia* spp. (Fig. 79)is composed of several different species present in the middle tidal zone and is the characteristic species. This zone extends from a plus 3 to 6 foot tidal level. Another spermatophyte *Jaumea carnosa* is present as patches in this zone. The principle animal present here as well as the upper and lower tidal zones is the California horn snail, *Cerithidea californica* (Fig. 137). Small polychaetes belonging to Family Spionidae (Figs. 63, 64) build tubes in the sediment.

The lower zone, which extends from a plus 2 to 4 foot tidal level, is characterized by the cord grass, *Spartina foliosa* (Fig. 579). Large areas of cord grass are present in upper Newport Bay especially in those areas separated from land by meandering channels. The California horn snail is present along with spionid polychaetes, as in the middle zone, and different species of clams, for example the jackknife clam, *Tagelus californianus* (Figs. 257, 258).

Mud flats and tidal channels occur below the lower zone which is generally absent of grasses and flowering plants. During the spring and again in the fall extensive growths of *Enteromorpha crinita* (Fig. 525) and other species of the genus as well as the sea lettuce, *Ulva lobata* (Fig. 527) occur on the mud flats. The California horn snail reaches its population peak on these mud flats. Many species of shore birds feed on the animals present on and in mud during low tides. A person should be extremely cautious about walking onto the mud flats since it is possible to sink into the mud above the knees.

Intertidal zonation of a sandy to muddy beaches is more easily seen within the protected waters of a bay because of less substrate movement during a tidal cycle. In such areas, holes in the substrate are seen especially in the mid-tide to minus tidal zone. These holes are made by, but not limited to, such animals as the pelecypods *Saxidomus nuttalli* (Figs. 233, 234), *Macoma nasuta* (Figs. 243, 244), *Tagelus californianus* (Figs. 257, 258), the ghost shrimp *Callianassa californiensis* (Fig. 323) and by many species of polychaetes. Polychaete tubes, such as by members of the Family Onuphidae (Fig. 57) or the Family

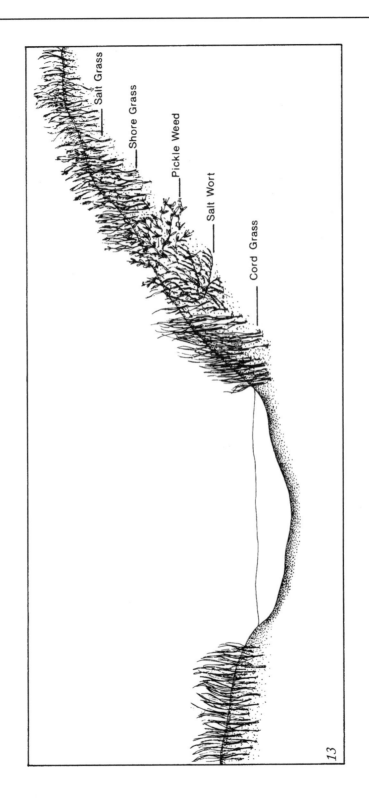

Salt Grass

Shore Grass

Pickle Weed

Salt Wort

Cord Grass

13

Spionidae (Figs. 63, 64) can be seen projection above the surface of the substrate. Many species of snails can been seen crawling over the surface of the muddy sand flats during low tides. Plants such as the eel grass *Zostera marina* (Fig. 574) are attached by their roots in the lower tidal zone in these protected waters.

PLANKTON

The word plankton means floating and refers to floating organisms in biology. Planktonic organisms have not been emphasized herein largely because they are generally microscopic in size. Planktonic organisms are generally collected by means of a cone-shaped net with a jar at the end. The net is made of cloth of a fine weave with minute openings. Plankton is collected by hauling the net through the water from a boat, throwing it into the water and bringing back, or walking on a dock and trailing the net in the water. The water passes through the openings in the cloth leaving organism trapped within the net which are gradually washed into the jar at the end of the cone. Commercial nets are available but such nets can be made at home with fine nylon netting. The most frequently encountered planktonic organisms include diatoms (Figs. 522-524), dinoflagellates, such as *Gonyaulax polyhedra* (Fig. 15), copepods, such as *Calanus* sp. (Fig. 277), arrow worms, such as *Sagitta euneritica* (Fig. 377), and larval stages of invertebrates and fish. Identification of planktonic organisms requires the use of a microscope.

KELP FOREST

The dominant species of the kelp forest is the giant brown alga *Macrocystis pyrifera* (Figs. 14, 540). It attaches to a rocky substrate by means of a tangle of root-like structures known as holdfasts. Rising from the holdfasts are a number of stalks (stipes) which bear numerous gas-filled bladders (called pneumatocysts). These bladders give buoyancy to the stalk. Up to 40 stalks are known to arise from one holdfast. A single stalk may have as many as 240 blades which include the bladder. Maximum growth and density occurs in water depths of 25 to 60 feet. The stalk extends to the surface where the buoyant fronds spread horizontally for 20 feet or more. These fronds form a canopy, termed the kelp canopy, which like its terrestrial counterpart shades the regions below. *Macrocystis pyrifera* is regarded to be the largest marine alga in the world with lengths greater than 150 feet recorded. Growth rates are rapid with over one foot per day having been measured.

The giant kelp is known from Santa Cruz, California south to Punta Asuncion—Punta Hipolito, Baja California. In Los Angeles and Orange Counties it is present at Point Dume, Palos Verdes Peninsula and wherever offshore rocks occur in Orange County. The extent of kelp forests in Southern California vary from year to year. Many physical, biological and anthropogenic factors can affect the distribution of kelp beds.

While *Macrocystis* is the dominant alga present, there are many additional species present. Other larger species of brown algae which may be present include a related species *Macrocystis angustifolia*, *Egregia menziesii* (Figs. 542, 543), *Cystoseira osmundacea* (Fig. 547), and *Eisenia arborea* (Fig. 544). Many small species of red algae are also present.

The kelp forest community is home to hundreds of species of invertebrates and fish. While polychaetes, mollusks and crustaceans are the most frequent in terms of species and specimens, it is the sea urchin which is the most conspicuous invertebrate. The purple sea urchin, *Strongylocentrotus purpuratus* (Fig. 390), the red sea urchin, *Strongylocentrotus franciscana* (Fig. 389), and the white sea urchin, *Lytechius anamesus* (Fig. 391), are herbivores and can feed all along the stalk and are capable of destroying the plant. Numerous gastropod species also feed upon the kelp including such species at the sea hare, *Aphysia californica* (Fig. 190), Norris' top shell, *Norrisia norrisi* (Figs. 119, 120), the wavy top turban, *Astraca undosa* (Fig. 129) as well as various abalone species.

Many species of fish inhabit the kelp beds but none is completely dependent upon the kelp for survival. Most of the species of the kelp forest are browsers, plantivores or carnivores. The halfmoon, *Medialuna californiensis*, and the opaleye, *Girella nigricans* (Fig. 452) are the only species of fish that regularly feed upon the kelp and associated species of seaweed. The most colorful species of fish that inhabit the kelp bed is the orange colored garibaldi, *Hypsypops rubicunda* (Fig. 459).

Different species of birds and mammals are found in association with kelp beds where they feed upon the fish and larger invertebrates such as shrimp. None of these species of birds and mammals are completely dependent upon the kelp bed.

Kelp is harvested and processed for different chemicals. See the account of *Macrocystis pyrifera* for additional information about this species.

RED TIDES

Red tides occur sporadically in the marine waters of Southern California. The causative organism is usually the red dinoflagellate protozoan *Gonyaulax polyhedra* (Fig. 15). Red tides caused by this species usually occur during the summer and fall months, but spring red tides are the result of another dinoflagellate *Procentrum micans*. Dinoflagellates are microscopic in size, but when countless numbers of them are present, the water becomes red in color. The more northern species of *Gonyaulax*, *G. catenella*, which is the causative agent of paralytic shellfish poisoning, may cause some red tides in Southern California. *Gonyaulax polyhedra* form cysts (resting stage) which are found on the bottom, and an unknown triggering agent or agents stimulate the growth of this organism. This species has the characteristics of a plant in that it is able to manufacture its own food by the process of photosynthesis. This organism is also able to move about with its flagella and take in food as many microscopic animals. This species is bioluminescent which means it is capable giving off "living light" at night whenever it is disturbed. Whenever red tides occur near shore, it is possible to stand on a floating dock at night and observe fish swimming through the red water and leaving a trail of light behind.

The mussel quarantine, which extends from May 1 through October 31, should be observed because of the possibility that mussels or other clams may contain in their bodies the paralytic poison found in *Gonyaulax catenella*. Because mussels and clams filter water through their bodies to feed, they can concentrate millions these red tide organisms in their digestive tract at one time. If the red tide organism is *Gonyaulax catenella*, then it is possible for a person to contact paralytic shellfish poisoning. This poison, an unknown toxin, caises paralysis and even deaths of people along the Pacific Coast above Point Conception since recorded history. The California Department of Public Health has devised a test for measuring the toxic levels of this poison.

Red tides occur either offshore or in bays and harbors. Offshore blooms of red tides generally occur in streaks or patches and they can be easily seen from a fishing pier. No health hazard is known to occur to persons swimming in a red tide caused by either species of *Gonyaulax*. If the red tide occurs within a bay and if the conditions, whatever these may be, are ideal for *Gonyaulax polyhedra*, they will reproduce within the bay causing to a more intense red tide. Some of these organisms will begin to die as a normal process, but because of the intense population of them and because of limited water circulation within the bay, the die-off leads to a decomposition of these organisms. Decomposition requires dissolved oxygen. This leads to a reduction in the dissolved oxygen concentration in the water which, in turn, leads to the death of the more sensitive animals, such as fish, which further depletes the dissolved oxygen supply in the water. If the red tide is extensive, the net result may be a massive die-off of not only attached organisms but also the bottom dwelling animals as well. If such a die-off occurs it generally takes a few months for the particular area to recover to normal conditions.

MARINE BORERS

The principle marine wood borers found in Southern California are the gribble, *Limnoria tripunctata* (Figs. 299, 300) and the ship worm, *Lyrodus pedicellatus* (Figs. 270, 271). A third type, the amphipod *Chelura terebrans* (Fig. 305), is of lesser importance in terms of destruction in local waters. Another species of gribble, *Limnoria quadripunctata*, and two species of ship worms, *Teredo navalis* and *Bankia setacea*, are present in Southern California. They are found in fewer numbers in wooden structures in the colder, offshore waters. Attacks by the gribble *Limnoria tripunctata* have been observed in Marina del Rey, Los Angeles-Long Beach Harbors, Alamitos Bay, Anaheim Bay, Newport Bay, and offshore pilings in Santa Monica Bay. This species is capable of burrowing through creosoted pilings into the non-treated inner portion. Attacks by the ship worm, *Lyrodus pedicellatus*, are known to occur in Los Angeles-Long Beach Harbors, Alamitos Bay and Newport Bay. This species cannot burrow into creosoted pilings. The wood borer problem has been greater in Los Angeles-Long Beach Harbors in past years because of the greater number of wood pilings and wooden boat floats. In recent years the amount of destruction by wood borers has been lessen by the gradual conversion to concrete pilings and fiber glass boats and boat floats.

FOULING ORGANISMS

Fouling organisms are defined as plants and animals which attach to man-made structures. Depending upon the type of structure, fouling organisms can become a nuisance especially when they attach to boats, water intake pipes, etc. The speed of a boat or ship can be slowed by as much as 10-20% by a large quantity of fouling organisms.

A population of fouling organisms fluctuates considerable during the course of a year, but a well establish community of about 2 years of age or older is dominated by the bay mussel, *Mytilus edulis* (Figs. 11, 203, 204). The number of associated organisms vary considerably during the year with a greater variety of animals, especially polychaetes and amphipods, found during the months of July through September. Over 80% of the organ isms which settle on boats occur during these months.

Lowest diversity of marine life occurs during January and February. Many of the marine phyla are represented in a mussel bed attached to a floating dock; in fact, this is a very convenient source of living material for classroom use which is readily available without the concern of tidal conditions. Furthermore, since this is material attached to a structure made by man and is located in a highly modified marine environment, collection of this material for classroom use is not decimating natural marine areas.

MARINE FOOD CHAINS

The simplest aquatic food chain consists of a primary producer, a primary consumer and a secondary consumer. A primary producer is an organism which contains chlorophyll and is able to absorb sunlight energy and transform nutrients into organic compounds by a process termed photosynthesis. In the ocean the primary producer is primarily a diatom (Figs. 522-524). The primary producer is fed upon by a primary consumer which is usually termed a herbivore. Copepods (Fig. 277) are an example of a primary consumer. A secondary consumer, or carnivore, which may be a fish, feeds upon the copepod. However, most food chains in the marine environment are not this simple. Other organisms are involved in the chain. Frequently there are tertiary consumers and beyond which may involve larger fish, birds, marine mammals or humans. Bacteria play a dominant role in breaking down the complex organic molecules ;into simpler forms making them available to be recycled by the primary producer.

The example cited above is one which occurs in the pelagic marine environment; however, each habitat has a distinctive food chain. The rocky intertidal environment can be given as an example of another type of food chain. The primary producer is algae which are attached to the rocks (diatoms are generally attached to the fronds of the algae). The algae is fed upon, for example, by crustaceans and polychaetes. The fecal matter from these animals (detrital feeders) may be fed upon by ostracods, nematods or brittle stars which live within the fronds and holdfasts of the algae. These animals which inhabit intertidal algae are fed upon by browsing fish such as the opaleye. The opaleye is not a permanent resident of the intertidal algal community and it may be fed upon by larger fish or other vertebrates elsewhere.

Food chains can be determined for each unique habitat by keeping in mind the basic three step process including the role that bacteria play. By knowing which plants and animals are present in the habitat and the food habits of the animals, it is possible to construct a provisional food chain for that site. The food habits for each animal group and, in some cases, each species, are included in the text. This information will assist the reader in constructing a food chain for a particular niche or habitat.

CONSERVATION OF NATURAL RESOURCES

With the increased interest in the oceans and marine biology, with the increase in the amount of recreational time available, and especially with the concentration of people in Southern California, it is of vital importance for us to conserve and preserve the marine life and marine environment in the local marine waters, as well as elsewhere. These remarks apply equally well to rocky shores, sandy beach and to bays. All too frequently collections of larger marine animals, such as starfish, sea urchins, snails and crabs end in trash barrels after they begin to smell. It would have been better to have left the animal in the ocean for the next person to observe. Collection of marine life with a specific purpose in mind, such as additional study, is one thing, but collecting without any purpose in mind should be discouraged.

Some general rules should be remembered whenever one is collecting or observing marine organisms. Rocks, which have been turned over, should be replaced to their original position otherwise, the plants and animals which were originally on the upper surface are now on the bottom and will die; the same, in reverse, holds for the animals which were originally on the bottom of the rock. Whenever digging in the sand or mud for clams or other forms of burrowing animals, the material should be replaced, otherwise many of these animals will also die because their habitat was disturbed. Perhaps the single rule to remember with regards to the conservation of marine life is to be aware that each species has its own specific habitat and whenever its environment is disturbed, the chances are that the organism will die.

There are several marine aquaria in Southern California which offer an opportunity to see close up a large variety of marine life. The Cabrillo Beach Marine Aquarium in San Pedro features marine life of Southern California. The aquarium at Scripps Institution of Oceanography in La Jolla and Sea World in San Diego also feature local marine life as well as animals from elsewhere. Many large aquaria, such as the one in Monterey, California are found or planned for other parts of the country.

Many marine areas of Los Angeles—Orange Counties, especially rocky shores, have been set aside as marine biological preserves by the State of California with the California Department of Fish and Game as the regulatory agency. Marine preserves are posted with signs which state the regulations as follows:

> In accordance with the California Fish and Game regulations only abalone, mackerel, lobster, bonita, halibut, perch, and sand bass, spotted bass, kelp bass, croaker, corbina, and rockfish may be taken in the marine refuge. All other aquatic life is protected and may not be removed. Do not remove rocks or shells or other material from the tide pool environment. Fish and Game Code Section 10664.

The reason why the abalone and fish are excluded from protection in the regulation is that they are covered by other sections of the Fish and Game Code. The success or failure of marine preserves depends upon the individual citizen. Marine preserves can be made more beautiful by utilizing trash containers and not littering the beaches with cans, paper, bottles and plastic goods. Consult the California Department of Fish and Game for further information.

Back bay areas such as Bolsa Chica or Upper Newport Bay are important habitats for resident and migratory birds. Since much of the wetland area of Southern California has been altered, there have been a considerably amount of interest both by the private and public sector in restoration of some of these wetlands. Some of these restored areas include improving fish nurseries in Anaheim Bay (within the Naval Weapons Station of Seal Beach) and construction of islands for nesting sites for the Least Tern and other birds in Talbert Wetland (Huntington Beach) and Upper Newport Bay. Restoring marine areas is a new and evolving science both in California and elsewhere. In order for the restoration of the area to be successful, the entire ecosystem must be considered.

Endangered and Threatened Marine Animals of Southern California. The Endangered Species Act of 1973 and the amendments of 1982, enacted by the Congress of United States, provides protection to those species which are considered to have a population beneath its ability to sustain itself (endangered species) or considered to be at or near it sustainable limits (threatened species). The table below lists the marine animals of Southern California which are either endangered or threatened. The California Department of Fish and Game publishes a list each year of the threatened or endangered species in the State. There are many additional species of freshwater and terristrial animals which are either endangered or threatened which are not listed herein. There are also many plants which are either endangered or threatened, but none of them live in the ocean.

Endangered and Threatened Marine Animals of Southern California

Species	State		Federal		Figure
	Endangered	Threatened	Endangered	Threatened	Number
Birds					
Pelicanus occidentalus Brown pelican	X		X		486
Rallus longirostris Clapper rail	X		X		499
Charadriua alexandrinus Snowy plover				X	503
Sterna antillarum Least tern	X		X		493
Passerculus s. beldingi Belding's savannah sparrow	X				514
Mammals					
Arctocephalus townsendi Guadalupe fur seal		X		X	
Eumetopias jubatus Northern sea lion				X	
Enhydra lutris Southern Sea otter				X	519
Balaenoptera borealis Sei whale			X		
Balaonoptera musculus Blue whale			X		
Balaenoptera physalus Finback whale			X		
Megaptera novaeangliae Humpback whale			X		
Balaena glacialis Right whale			X		
Physeter catodon Sperm whale			X		

HOW TO USE THIS BOOK

Plants and animals are arranged into various groups termed phyla (singular: phylum) on the basis of their particular characteristics. Each phylum is subdivided into one or more classes; classes into one or more orders; orders into one or more families; families into one or more genera; and the genera into one or more species. The scientific name of an organism consists of a genus and species name. The name is latinized and is printed in italics. The rules for naming a new species of plant or animal is governed by precise procedures. Often times the abbreviation "sp." follows a generic name; this means that the specific name of the species is unknown. Common names are not governed by any rules and are derived from common usage by the layman. Common names are often coined from a unique feature of the organism, such as color, marking, habitat, etc. In this book the common name follows the scientific name. However, not all plants and animals have a common name. The common name often varies with the geographical region. In the example given below, *Mytilus edulis* is known locally as the bay mussel but elsewhere it is known as the blue mussel or the edible mussel. The system of classification can be summarized with the following example:

Phylum Mollusca
Class Bivalia
Order Filibranchiata
Family Mytilidae
Genus *Mytilus*
Species *edulis*
Common name: the bay mussel

This book is arranged according to the various phyla beginning with the one-celled animals (Phylum Protozoa) and completing with the marine grasses (Phylum Spermatophyta). Some distinguishing characteristics are given for each phylum, class and in some cases order and family. The characteristics used in this book are those features which can be seen without the use of a microscope unless specifically stated otherwise.

Each species is discussed according to its size, color, and distinguishing characteristic features in order to separate it from other species. The specific localities are given where either the author or his students have actually seen the organism. It should be emphasized that the species may occur elsewhere; the area may not have been studied as extensively; the species may have been overlooked; or seasonal or yearly fluctuations may have excluded the particular species at that time the area was visited. The inclusion of the specific locality is intended to further increase the usefulness of this book. Additional pertinent information is included as an aid to its identification. Notes on its biology are included, if known. Not all the marine plants and animals of Southern California have been included in this edition; only the more conspicuous forms of a particular phylum or representative species of a particular are included. Emphasis has been placed on those species which inhabit intertidal waters; however, some of the larger subtidal organism are represented. The index list alphabetically the group names, the scientific names, the common names and some additional information. The appendix includes a list a references which give additional information about the marine area of the Southern California Bight and its biota.

Three ways are suggested to facilitate the identification of a particular plant or animal. If the phylum or the name of the group is known, then the reader should turn to that section of the book (page references are given in both the Table of Contents and the Index). The reader should then refer to the figures within that particular section and attempt to identify the organism by matching the specimen with the figure. The discussion included with each figure should be additional assistance in identifying the organism

If, on the other hand, the phylum to which the organism is unknown, then the reader could thumb through the book and attempt to match the specimen with the figure. Initially this method of identification by leafing through the pages of the book will be slow and is dependent upon the previous knowledge by the reader of biology. To readers with a limited previous experience with marine biology, the author suggests working with snails and clams first as a means of gaining experience.

A third method of identifying the organism is by using the key given below. While the use of the key will not lead you the name of the organism, it will assist the reader in determining which group of organism the specimen belongs. Once the phylum or group is known, the reader can follow the first procedure described above.

To use a key to identify the group to which your organism belongs begin with couplet la and 1b. Chose one of the contrasting statements and proceed to the couplet number indicated at the end of the chosen statement. Continue this selection until the name of the organism is cited instead of a couplete number then follow the first procedure described above.

Key to the Major Groups of Marine Plants and Animals
from Southern California

la.   Organism microscopic, requires the use of a dissecting or compound microscope to see 2

1b.   Organism macroscopic, easily seen with an unaided eye 4

Key to Microscopic Organisms

2a.   Organism a single cell or a colony of relatively independent cells (Fig. 15) 3

2b.   Organism composed of a small number of cells; parasitic in cephalopods or other animals (Fig. 16). . . . . Mesozoa

3a.   Rigid cell wall, composed of silicon dioxide; geometric in appearance (Figs. 522-524) . . . . Chrysophyta (Diatoms)

3b.   Cell membrane (cell wall) may or may not be rigid; cell wall not composed of silicon dioxide (Fig. 15) . . . . Protozoa

Key to Macroscopic Plants and Animals

4a.   Organism plant-like in appearance; generally colored some shade of green, brown or red 5

| 4b. | Organism usually animal-like in appearance although some may resemble plants (see Figs. 24, 360). . . . . 8 |

Key to Macroscopic Plants

5a.	Plant structure usually includes roots, stems, leaves and flowers (may not be present at the time); generally present in wetlands or intertidal zone (Fig. 572) . . . . Spermatophyta
5b.	Plant structure simple without roots (root-like structures may be present, see Fig. 544), stems, leaves or flowers. 6
6a.	Plants usually bright green to olive green in color (Figure 526) . . . . . . Chlorophyta
6b.	Plants may be olive green, tan, brown or various shades of red 7
7a.	Plants predominately tan or brown in color sometimes tinged with green; plants often large in size (1 foot or more in length) (Fig. 540) . . . . . . Phaeophyta
7b.	Plants frequently finely branched and generally some shade of red; rarely over 1 foot in length (Fig. 553) . . . . Rhodophyta

Key to Macroscopic Animals

8a.	Body branches resembling a plant (Figs. 24, 360). 9
8b.	Body not branched 10
9a.	Ends of branches with a circle of tentacles visible to the unaided eye (Fig. 35). . . . . Cnidaria, in part
9b.	Tentacles microscopic; body rigid and may be composed of calcium carbonate (Figs. 364-366) . . . . Ectoprocta, in part; Entoprocta
10a.	Body growing on the surface of substrate; animal often colonial 11
10b.	Animal otherwise 13
11a.	Irregular arrangement of openings of various sizes; openings scattered over the surface; spongy to touch (Fig. 17) . . . . Porifera
11b.	Openings, if present, microscopic; body may be fleshy but not spongy in texture 12
12a.	Body calcareous (Figs. 364-366). . . . . Ectoprocta, in part
12b.	Body fleshy which may be supported by a stalk; may be brightly colored (Figs. 413, 414). . . . . Tunicata
13a.	Radial symmetry with body shape arranged in a circle (Fig. 384). 14
13b.	Symmetry, if present, not radially arranged. 15
14a.	Radial symmetry typically in fives; body with short to long spines (Fig. 384). . . . . Echinodermata, in part
14b.	Radial symmetry with a circle of tentacles surrounding an opening (mouth); body usually fleshy (Fig. 35) . . . . Cnidaria, in part
15a.	Body worm-like in shape 16
15b.	Body not worm-like in shape 25
16a	Worm segmented (Fig. 52). 17
16b.	Body not divided into segments 18

| 17a. | Body generally less than an inch in length; mouth provided with a pair of chitinous jaws.................... Insect larvae (not included) |

17a. Body generally less than an inch in length; mouth provided
   with a pair of chitinous jaws.................... Insect larvae (not included)
17b. Segmentation distinct which may (Fig. 52) or may not have (Fig. 68)
   lateral projections; anterior end may be provided with tentacles
   (Fig. 78) .... Annelida, Polychaeta
18a. Body flattened.. 19
18b. Body rounded not flattened (Fig. 86) 20
19a. Body oval in shape; tan to brown in color (Fig. 41) .... Platyhelminthes
19b. Body narrow, colorless; spines at anterior end (Fig. 377) .... Chaetognatha
20a. Body fleshy, resembling a sausage or peanut in shape.................... 21
20b. Body shape otherwise 23
21a. Tentacles present at one end of body which may be withdrawn
   into the body. 22
21b. Tentacles absent (Fig. 86)................................... Echiura
22a. Body surface smooth (Fig. 88) .... Sipuncula
22b. Body surface smooth or with bumps; muscular layers in
   fives (Fig. 401).. Holothuroidea
23a. Tentacles present (Fig. 376) .... Phoronidea
23b. Tentacles absent 24
24a. Soft-bodied worms lacking any additional structure (Fig. 42) .... Nemertea
24b. Body divided into three regions (Fig. 418) .... Hemichordata
25a. Body colorless, gelatinous; olive-like in shape (Fig. 40). .... Ctenophora
25b. Body shape and color otherwise 26
26a. Body segmented and often in 2 or 3 distinct region, covered with a
   hard skeleton; with jointed appendages....................... Arthropoda
26b. Body not segmented; appendages, if present, not jointed. 27
   (Note 4 choices for number 27)
27a. Body covered or enclosed in 1, 2 or 8 calcareous shells
   (Figs. 90, 137, 201, 202, 275) .... Mollusca, in part; Brachiopoda
27b. Body fleshy, slug-like in appearance (Fig. 199)....... Mollusca, Opistobranchiata
27c. Body with 8 or 10 tentacles with suckers (Fig. 273) ..... Mollusca, Cephalopoda
27d. Body otherwise 28
28a. Animal 1-2 inches in length; fish-like in appearance
   (Fig. 417).. Cephalochordata
28b. Animal with backbone; body covered with scales, feathers or hair
   (Figs. 433, 486, 517)................................... Vertebratae

MARINE ANIMALS

PHYLUM PROTOZOA
The one-celled animals

Protozoans are one-celled animals which live in all environments of the world wherever sufficient moisture is present. They are small in size and require the magnification of a compound microscope in order to be seen. Protozoans move by action of flagella, cilia or by pseudopodia while some are unable to move. Protozoans feed on bacteria, organic debris, each other or they manufacture their own food by the process of photosynthesis. In local marine waters protozoans can be found in the water mass, on the bottom or attached to other organisms.

Gonyaulax polyhedra (Fig. 15). This species of protozoan moves by the action of two flagella; one is equatorial, the other extends from the posterior end. The body is covered with cellulose plates and is referred to as an armored dinoflagellate. *Gonyaulax polyhedra* lives in the water mass especially near the surface where it carries out photosynthesis. When this organism is present in countless numbers, the water may become red in color causing red tides. See the section on red tides in the introduction for further discussion of this phenomenon.

PHYLUM MESOZOA

Mesozoans are parasitic animals which inhabit body cavities and organs of octopus, squid, brittle stars, flatworms, clams, polychaetes and other animals. They are small in size with the body composed of 25 or less cells. They have complex life cycles, many of which are incompletely known. Mesozoans lack a digestive tract and therefore obtain their nurishment directly through the cell membranes. In order to study this group it is necessary to make smears of tissue on slides, fix with Bouins, and stain with hemotoxylin and eosin. It is necessary to examine the slides under a compound microscope in order to see these animals.

Dicyemennea abelis (Fig. 16). This dicyemid mesozoan is found in the kidneys of nearly all specimens of the common intertidal octopus, *Octopus bimaculatus* (Fig.231). It apparently

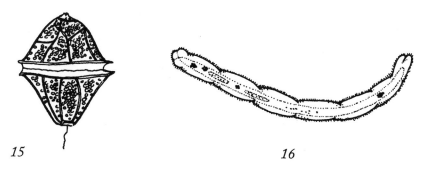

15   16

does little or no damage to its host. The complete life cycle of this species, as in the case with all dicyemid mesozoans, is unknown; presumably, an unknown stage occurs outside the octopus, possibly in another host.

PHYLUM PORIFERA
The Sponges

The members of the Phylum Porifera are known commonly as the sponges or pore-bearing animals. They pump water through microscopic openings over the body and discharge this water through one or more large openings or oscula at the top (Figs. 17, 18, 20). The water brings microscopic food (bacteria and fine particulate matter) and dissolved oxygen into the body and carries out waste products and carbon dioxide through the osculum. A compound microscope and chemicals are necessary in order to identify sponges. Sponges secrete microscopic structures known as spicules, which may be either calcareous or siliceous in chemical composition. The chemical nature of the spicules is tested by the addition of a small amount of acetic acid on a piece of the sponge. If bubbling occurs, this indicates the presence of calcareous spicules. The shape of the sponge varies considerably depending upon the competition for space with mussels and other organisms. Sponges are commonly encountered in the intertidal waters of southern California. Many species may be seen attached to rocks but because of the difficulty in identification only a few distinctive species are discussed herein.

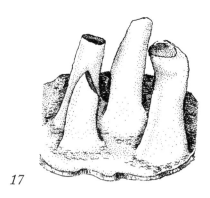

17

Haliclona permollis—the purple sponge (Fig.17). This sponge is light purple in color. It may become several inches in diameter at its base and extend outward 2-3 inches. The body consists of many finger-like projections, each of which terminates with an osculum. The spicules are siliceous in nature and appear as simple rods in a netlike arrangement. This species has been found in protected waters attached to boat floats in Marina del Rey, King's Harbor, outer Los Angeles-Long Beach Harbors, Alamitos Bay and Newport Bay. The largest specimens generally occur during the fall months.

18

Leucosolenia sp.—the white sponge (Fig.18). The general body form and size is similar to *Haliclona permollis*; however, they differ in color and by the structure of their spicules. *Leucosolenia* sp. is composed of calcareous, 3-pointed spicules. The occurrence of this species in protected waters is similar to *Haliclona permollis*. It is encountered more frequently in summer and fall.

Plocamia karykina—the red sponge (Fig. 19). This species may be identified by its size, color and habitat. It reaches an inch in diameter, is red in color and is found attached to rocks, especially on the undersurface of small ledges in the intertidal environment. This species has been collected from several localities of the Palos Verdes Peninsula, Little Corona, Laguna Beach, and Dana Pt.

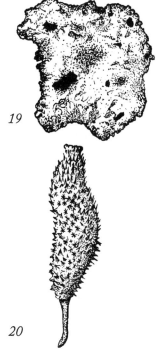

19

20

Vase sponge (Fig. 20). Undoubtedly 2 or more species of vase sponges occur in southern California waters. All are white or light gray in color and about 1/2 inch in length. Spicules penetrate the body and also form a circle around the osculum as shown in Fig.16. This type of species has been collected from boat floats in Marina del Rey, Alamitos Bay and Newport Bay.

PHYLUM CNIDARIA
The jellyfish, hydroids, corals, sea anemones, etc.

Cnidarians, formerly known as coelenterates, may be solitary or colonial animals which frequently resemble plants. They are characterized by the presence of nematocysts or stinging cells. In some parts of the world, generally tropical areas, these stinging cells may be dangerous or even result in death to man; none of the local species is particularly dangerous although some jellyfish may cause some discomfort to a sensitive person. Cnidarians possess radial symmetry which is best shown by the sea anemone (Fig. 35). These animals have a mouth and digestive system but lack an anus.

CLASS HYDROZOA—The hydroids (Figs. 21-30). The majority of the species possess a conspicuous colonial plant-like or hydroid stage (Figs. 21-24, 26-28) and a small solitary jellyfish or medusoid stage. The medusae produce the hydroid stage as the result of the fertilization of an egg and sperm. The hydroid stage is composed of one or more individuals (polyps) which feed upon microscopic life; the polyps are interconnected so that what one polyp eats is distributed to the entire colony. The medusoid is either a non-feeding stage or it feeds upon particulate matter.

Aglaophenia sp.—the ostrich plume (Fig. 21). This colonial hydroid is gray to brown in color and may reach 6 inches in length. It is easily recognized by its feathery appearance. Several species are known from Southern California but specific determination is difficult. The medusoid stage is reduced and is not a free-living form. The ostrich plume is more frequently encountered on sandy beaches where specimens are attached to windrows of different species of algae. Living specimens can be collected from the low intertidal horizon

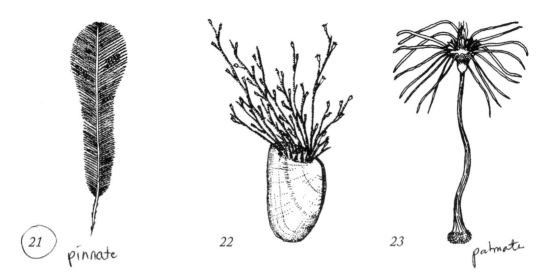

21 pinnate   22   23 palmate

where they are attached to the stipes (stalks) of brown algae. Clumps can be collected from the boat floats near the entrance of Alamitos Bay and Newport Bay especially where there is good water circulation.

Clytia bakeri (Fig. 22). This small species, which measures about one inch long, usually attaches to the bean clam, *Donax gouldi* (Figs. 252, 253) and less frequently to the Pismo clam, *Tivela stultorum* (Figs. 235, 236), the olive shell, *Olivella biplicata* (Fig. 181) and the basket shell, *Nassarius fossatus* (Fig. 174). These animals live just beneath the surface of an intertidal sandy beach. The hydroid extends above the surface of the sand. Large populations of *Clytia bakeri* may be present during a population explosion of the bean clam which occurs every 2 or 3 years in Southern California sandy beaches.

Corymorpha palma (Fig. 23). This white hydroid, which measures about 3 inches in length, resembles a small palm tree. It anchors itself in the substrate with root-like structures and extends above the surface. The medusoid stage is reduced and remains attached to the hydroid stage. Occasionally it can be seen during an extremely low tide in Alamitos Bay and Newport Bay, but greater numbers are known subtidally from these protected waters.

Obelia sp. (Figs. 24-25). The hydroid stage attaches to rocks, pilings, algae or sessile animals. Colonies (Fig.24) measure up to several inches in length. The reproductive polyp gives rise to the medusae (Fig.25) which measure up to 1/2 inch in diameter. The medusae may be taken in plankton tows, but generally they can be seen swimming in a dish of sea water containing the hydroid stage. Many species of *Obelia* are known from Southern California, but specific identification is difficult. Large specimens of *Obelia* have been collected from floats and pilings in Marina del Rey, Los Angeles Harbor, Alamitos Bay

24   Palmate

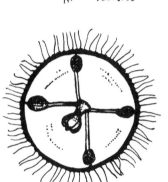

25

and Newport Bay especially during the spring months. This genus is typically used as an example of a hydrozoan in biology classes. Smaller colonies, measuring less than an inch, have been taken from most of the rocky shores during minus tides; they are attached to algae or rocks.

Sertularia sp. (Fig.26). This hydroid is tan colored, stiff and measures up to one inch in length. It can be distinguished from *Obelia* by the alternating arrangement of the polyps. The medusoid stage remains within the reproduction polyp of the hydroid where fertilization occurs. *Sertularia* sp. is found at the rocky shores attached to algae and rocks in the low tide horizon.

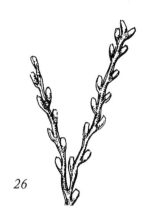

26

Tubularia sp.—the naked hydroid (Figs. 27-28). Large gray to pink colonies, measuring 3-4 inches in length, attach to floating docks or submerged structures wherever there is sufficient current. Under ideal conditions extensive masses can develop which will harbor many other invertebrate animals such as flatworms, polychaetes, nudibranchs and amphipods. The medusoid stage is reduced;

palmate —

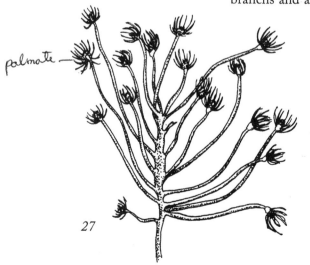

27

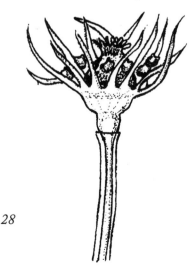

28

fertilization and development of the larvae takes place within the hydroid stage. The actinula larvae are released from the hydroid stage after several minutes in the laboratory. *Tubularia* has been collected from boat floats in Marina del Rey, Los Angeles-Long Beach Harbors, Alamitos Bay and Newport Bay.

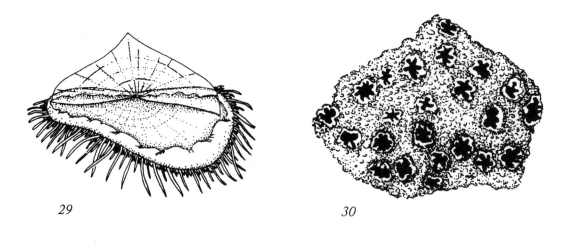

29   30

Velella velella—by the wind sailor or purple sailor (Fig. 29). This purple hydroid, which measures 2-4 inches in diameter, floats on the surface of the ocean. The projecting portion acts as a sail, hence the common names. The appearance of this species is erratic, but it is more frequently encountered during the late spring months. In some years countless numbers of these may be washed ashore, other years, none. The specific causes of its appearance are unknown, but presumably some particular oceanographic condition(s) triggers a reproductive bloom. This species was once thought to be related to the Portugese man-of-war found in marine waters, but it proved to be only a superficial resemblance. Its stinging cells lack the strong toxins found in the Portugese man-of-war.

Allopora californica—(Fig. 30). This colonial animal resembles a staghorn coral of the tropical seas; it grows to 1 foot in length. It is purple to pink red in color. It only occurs in subtidal waters especially on the vertical faces of rocks where the currents are strong.

CLASS SCYPHOZOA—the jellyfish (Figs. 31-32). This class of enidarians contains the majority of the jellyfish. These animals are generally transparent and measure less than 2-3 inches in diameter. Some species are vividly colored and measure more than one foot in diameter. These animals are pelagic and are generally found at or near the surface; they are frequently washed up on the beach where they soon dry out.

Aurelia aurita (Fig. 31). This transparent species measures about 3 inches in diameter. It is distinguished by its flattened umbrellar shape. The occurrence of this jellyfish is erratic;

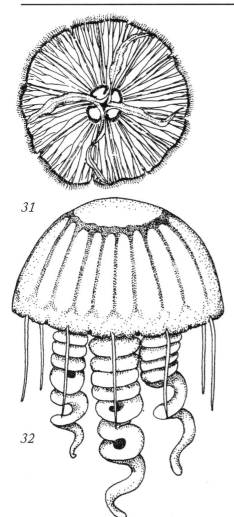

31

32

it can be seen on occasion in the spring months floating in large numbers near the entrances of any of the local bays or harbors.

This species is often used in biology classes as a representative of this group.

Pelagia panopyra—**the purple striped jellyfish** (Fig. 32). This jellyfish is characterized by its large size, one foot in diameter, and its purple stripes against a transparent background. It is present nearly every year during the late spring-early summer months between the mainland and the offshore channel islands. Occasional specimens are washed upon the beaches. The stinging cells even from a washed up specimen on the beach may be painful to some people but generally the pain is gone within minutes or an hour or two. The pain can be reduced by washing the affected area with a dilute solution of household ammonia or hot water.

CLASS ANTHOZOA—the sea anemones, sea pansies, sea pens and corals (Figs. 33-39).This group of animals is typically brightly colored and superficially resemble flowers, hence the common names on many of the species.

Subclass Alcyonaria (Figs. 33, 34). These animals possess pinnate tentacles which distinguishes them from the Anthozoa (below).

Renilla kollikeri—**the sea pansy** (Fig. 33). The sea pansy is purple in color and measures up to 3 inches in diameter. The stem (or peduncle) extends into the sandy sediments and the flattened portion (or rachis) lies on top of the sediments. The upper surface of the rachis contains many white colored polyps, each with 8 pinnate tentacles, which feed upon microplankton. Living specimens exhibit luminescence when disturbed after having been kept in the dark for several hours. Occasional

33

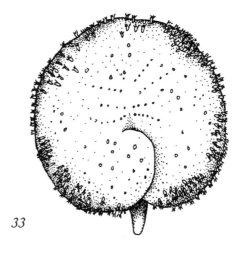

specimens may be taken at low tide in Newport Bay, but unfortunately these specimens are generally quickly collected by the laymen. The sand star *Astropecten armatus* (Fig. 379) feeds upon the sea pansy. Populations of the sea pansy can be observed subtidally in sandy sediments.

Stylatula elongata—**the sea pen** (Fig. 34). This animal is tan in color and shaped like a pen. It measures up to a foot in length. The sea pen projects above the substrate and it is capable of moving up or down in its burrow or even bury itself. It is bioluminscent. This species is rarely seen intertidally. It has been collected near the entrances of Marina del Rey, Los Angeles-Long Beach Harbors, Alamitos Bay and Newport Bay. Populations also occur in *34* subtidal offshore waters.

SUBCLASS ZOANTHARIA—**the sea anemones and corals** (Figs. 35-39). These animals have simple tentacles. Most of the zoantharians in Southern California are sea anemones, but some corals exist in offshore waters but they are not reef-forming species.

Anthopleura elegantissima—**the aggregating sea anemone** (Fig. 35). The column is generally green to white in color and the tentacles tipped with pink. The green color is due to the presence of green pigment and to symbiotic green algae living with the tissues of the sea anemone. Solitary specimens will reach 3 inches in diameter when expanded; whereas, aggregated specimens seldom reach one-half this size. This carnivorous species feeds upon some crustaceans and other small organisms and larvae which come in contact with its tentacles. Aggregated populations of this species are abundant on rock surfaces in the upper mid-tide horizon at Pt. Dume, Palos Verdes Peninsula, Little Corona, Laguna Beach and Dana Pt. These specimens generally have large quantities of shell fragments present which are attached to adhesive organs along the column which give the animal protection from the sun. Solitary specimens have been observed in the quiet waters of Marina del Rey, Los Angeles-Long Beach Harbors, Alamitos Bay and Newport Bay.

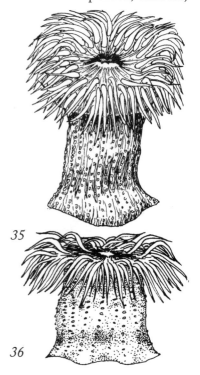

35

Anthopleura xanthogrammica—**the green sea anemone** (Fig. 36). This sea anemone is one of the most beautiful of our local intertidal invertebrates. When it is exposed to sunshine, it is deep green in color due to the presence of one-celled green algae in its tissue. When it is found in rock crevices or caves, it is pale green in color. This species may reach 6 inches in diameter and one foot in length *36* when expanded. It will feed upon dislodged mussels, crabs, sea urchins and other animals which come in contact with its tentacles. The green anemone is present in the mid-tide to low tide horizon at all rocky beaches and the outer reaches of rock jetties.

Diadume leucolena (Fig. 37). This small sea anemone measures less than one inch in diameter when fully expanded. The tentacles are pale salmon in color and the column is olive green with a few orange-yellow vertical stripes. It is found, often abundantly, in the quieter waters of Marina del Rey, Los Angeles-Long Beach Harbors, Alamitos Bay, San Gabriel River, Anaheim Bay, Huntington Harbour and Newport Bay attached to boat floats or pilings.

37

Corynactis californica—**the strawberry anemone** (Fig. 38). This brightly colored anemone is usually red, crimson or pink in color, but other colors occur. While it may appear singly, it is usually present as a colony of which all individuals are of the same color as a result of asexual reproduction by longitudinal fission. It is found in subtidal waters in Southern California attached to rocks or offshore pilings. (Fig. 12).

38

Moricea californica—**California golden gorgonean** (Fig. 39). Specimens are generally 1 to 2 feet in height, but some are known to reach 3 feet. The colony is supported by a central skeleton composed of a hard protein. Age of the colony can be determined by counting the annual rings in a cross-section of the skeleton. The numerous polyps, which are present on the branches, capture small animals or particulate matter for food. It is found in subtidal waters of Southern California attached to rocks or offshore structures.

39

PHYLUM CTENOPHORA
The comb jellies

The ctenophores are an entirely marine group of animals in which nearly all species are pelagic. They are called comb jellies because of the presence of eight

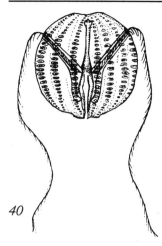

longitudinal rows of ciliated plates over a transparent jelly-like body.

Pleurobrachia bachei—the comb jelly or cat's eye (Fig. 40). This species is transparent and jelly-like in appearance. The body measures from 1/4 to 1/2 inch in diameter. Two long tentacles arise from either side which are used to capture small small crustaceans and other small animals from the water column. The area beneath the comb plates is bioluminescent in living specimens. *Pleurobrachia bachei* is usually collected with a plankton net. Its occurrence is sporadic in southern California waters. Occasional specimens are seen within the entrances of local bays and harbors.

40

PHYLUM PLATYHELMINTHES
The flatworms

This group of animals, which is characterized by having flattened bodies, includes the free-living turbellarians (Fig. 41) and the parasitic flukes and tapeworms. The parasitic species are known to occur on or in the local fish, birds, and mammals, but because they are not likely to be encountered by most people, they are not discussed herein. Identification of the free-living flatworms is difficult for the nonspecialist. Many species occur in Southern California, most of them are gray, tan or brown in color and measure less than one inch in length. They are carnivorous and feed upon small crustaceans such as isopods and am-

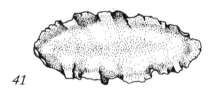

phipods as well as other small invertebrates. They are present in all local environments, except sandy beaches, wherever protection is available. Generally one or more specimens can be observed gliding over the undersurface of a rock overturned in the intertidal zone.

41

PHYLUM NEMERTEA
The ribbon worms

Members of this phylum are elongated, non-segmented worms which may be variously colored. Several species occur along the rocky shores especially within mussel beds or in clumps of algae or algal holdfasts. Two of the more commonly encountered species are described below.

Emplectonema gracile (Fig. 42). This worm will measure up to 10 inches in length. The upper surface is dark green and the lower surface is pale green to white in color. This species has been taken from mussel beds formed by *Mytilus californianus* (Figs. 201, 202) on the rocky shores and by *Mytilus edulis* (Fig. 203, 204) in the outer reaches of bays and harbors.

42   43

Cerebratulus californiensis (Fig. 43).   This worm may reach 1 foot in length and upwards to 1/2 inches wide. It is usually orange or reddish orange in color. It fragments easily into two or more pieces when collecting. It is carnivorous as are all nemerteans. It has been collected during low tides at Alamitos Bay and Newport Bay. It is frequently taken in subtidal benthic samples throughout Southern California from fine sediments.

PHYLUM ENTOPROCTA

The entoprocts are an obscure group of colonial animals which are almost all marine species. Each individual has a circle of tentacles with which it sifts out fine particulate matter from the water for food. The anus is located within the circle of tentacles which is the basis for the phylum name.

Barentsia sp. (Fig. 44).   Colonies of this species attach to rocks in the intertidal zone. Each individual measures about 1/2 inch in length. In the field colonies can be recognized by seeing their stalks swaying to and fro. Undoubtedly, this species can be found at all the rocky shores of Southern California; it has specifically been collected at Pt. Fermin and Little Corona.   *44*

PHYLUM ANNELIDA
The sea worms, earthworms, and leeches

Members of annelids are called segmented worms which refers to the numerous rings which encircle the body throughout its length. The annelids are divided into the classes Polychaeta, Oligochaeta and Hirudinea. The polychaetes, or sea worms, are largely a marine group and are an important component of the animal life in the sea. Oligochaetes are largely fresh water or terrestrial inhabitants although some small specimens can be collected from back bay areas of Alamitos Bay, Anaheim Bay or Newport Bay. Hirudinea or leeches are more frequently encountered in fresh water or tropical forests; they are either temporary parasites of vertebrates or carnivores. Marine leeches are parasitic on fishes. Neither the oligochaetes nor the leeches will be discussed herein.

CLASS POLYCHAETA—the sea worms. The majority of the species possess many bristles, or setae, on each segment of the body. These setae may arise directly from the body (Figs. 59, 60, 60, 68) or from lateral projections from the body (parapodia) (Figs. 45-58). Some of the species move about freely in their environment (for example, those species in Figs. 45, 48, 53, 58) while others are tube dwellers (for example, those species in Figs. 57, 63, 64, 71, 73-85). There are probably more than 200 species which occur intertidally in Southern California with many more present subtidally.

45

Family Polynoidae—the scale worms (Fig. 45). These worms are so named because the upper surface is covered with overlapping scales which number from 12-18 or more pairs. They are generally 2 inches in length and flattened. *Halosydna brevisetosa* (also known as *Halosydna johnsoni*) (Fig. 45) is gray in color, has 18 pairs of scales and measures about 2 inches in length. It is found in a variety of habitats in Southern California. It has been collected from all the rocky shores from the *Mytilus californianus* beds (Figs 201, 202) or on the undersurface of larger rocks in low tide zones. It feeds on small crustaceans and probably other small animals. In the laboratory it has grown larger on a diet of frozen brine shrimp. The largest specimens and greater numbers can be collected during the warmer months from clumps of *Mytilus edulis* (Figs. 203, 204) attached to boat floats in all bays and harbors of Southern California. Another scale worm, *Harmothoe lunulata*, has 15 pairs of scales, gray-brown in color and measures up to 1 1/2 inches in length. It inhabits the burrows of the ghost shrimp *Callianassa californiensis* (Fig. 323) which is found intertidally and subtidally in the local bays and harbors.

Family Sigalionidae (Fig. 46). This family also possesses overlapping scales on its upper surface but there are generally more pairs present than the species in Family Polynoidae. *Sthenelais fusca* is the most frequently encountered species in this family. It measures 4-5 inches in length and the body, including the scales, is rusty brown in color. Occasional specimens have been taken intertidally from the roots of surf grass (Fig. 441) along the Palos Verdes Coast.

Family Phyllodocidae (Fig. 47). The phyllodocids are often brilliantly iridescent in color and secrete large amounts of mucous. They possess either 4 (Fig. 47) or 5

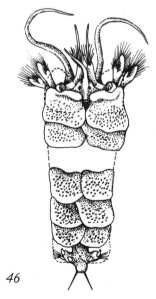

46

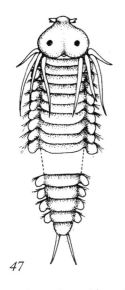

47

tentacles on their head and their parapodia appear leaf-like. Many species are known from Southern California, the majority of which cannot be identified in the field. *Anaitides medipapillata* (Fig. 47) appears iridescent purplish-brown in color and measures up to 6 inches in length. It has been collected from the different rocky shores of Palos Verdes Peninsula and Little Corona from under rocks or in mussel beds. It has been taken from the rock jetties and boat docks in Alamitos Bay and Newport Bay. Other species of phyllodicids can be taken from mussel beds and algal holdfasts; they generally measure about 1 inch in length.

Family Hesionidae (Fig. 48). These possess 2-3 tentacles and a pair of palps at the anterior end. *Ophiodromus pugettensis* (Fig. 48) may be either light in color (generally the free-living ones) or dark in color (commensal ones). The free-living population is found more frequently in all local bays and harbors on boat floats or pilings but an occasional specimen has collected from among the rocks at low tide at Pt. Fermin, Little Corona and Dana Pt. More people will encounter this species as a commensal on the sea bat starfish *Patiria miniata* (Fig. 382) which may be found intertidally under rocks at low tide at all rocky shores and rock jetties. Many individuals may be present on the undersurface of the sea bat especially during the winter months. Field studies have indicated that the commensal population is a changing one from day to day. For example, marked starfish may have 5 worms present one day but only 3 the next then 5 or 6 the third day. Presumably, the worms live in the sediment and move on and off the starfish. Both the free living and commensal populations are omnivorous feeders. A second species of hesionid, *Amphiduros pacificus*, resembles *Ophiodromus* but it is white to

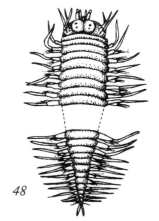

48

tan in color, slightly longer with specimens measuring 1 1/2 inches and with longer parapodia. This species is known from the mussel beds at Marina del Rey, Los Angeles Harbor, Alamitos Bay and Newport Bay.

Family Syllidae (Fig. 49-51). Specimens of this family are small with the majority measuring less than one inch. The prostomium generally has 4 eyes and 3 tentacles. The tentacles may be bead-like (Fig. 51), smooth (Fig. 50) or reduced in length (Fig. 49). The three general types of syllids are represented herein. *Exogone* sp. is white in color, measures less than 1/2 inch in length, has reduced tentacles and parapodial lobes and has been collected from the sediments of Alamitos Bay, Anaheim Bay and Newport Bay. *Odontosyllis phosphorea* (Fig. 50) is darkly pigmented and measures up to 2 inches. It is bioluminescent during its reproductive period which accounts for its specific name. It has been collected from among rocks in the low tide zone at several localities of Palos Verdes Peninsula, Little

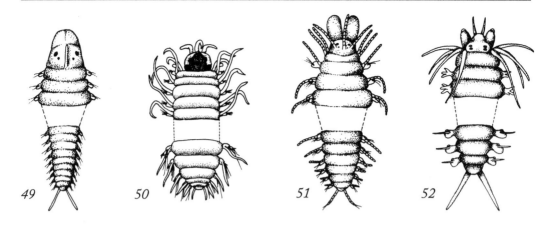

49 *50* *51* *52*

Corona, as well as the rock jetty of Alamitos Bay. *Typosyllis variegata* (Fig. 51) is one of several species present in Southern California which possesses beaded tentacles and cirri. It is light in color, measures up to 1 1/2 inches and has been collected from algal holdfasts at Lunada Bay and Little Corona.

Family Nereidae (Fig. 52). Many species of this family have been collected from a variety of habitats and they appear much like the species shown in Fig. 52; examination under the microscope is necessary in order to distinguish the different species. *Platynereis bicanalicu-lata* (Fig. 52), *Nereis grubei* and *Nereis latescens* all measure about 3 inches in length, green to tan in color and construct mucoid tubes in the sea lettuce *Ulva lobata* (Fig. 527), *Gigartina canaliculata* and other species of seaweeds. These species have been collected from boat floats in Marina del Rey, King's Harbor, Los Angeles-Long Beach Harbors, Alamitos Bay, Huntington Harbour, Newport Bay and from various species of finer branched specimens of red algae in the low tide horizon along all rocky shores in Southern California. *Neanthes succinea* is tan colored, measures 4 inches and has been collected from the mud flat areas of Alamitos Bay and Newport Bay. This species also occurs in Salton Sea where it was accidently introduced in the early 1930s. A few specimens of *Neanthes brandti* have been collected from the mud flats of Alamitos Bay; while specimens are known to measure 40 inches in length in the Pacific Northwest, these were only 6 inches in length. *Neanthes arenaceodentata* measures about 1-2 inches and has been collected from boat docks in Los Angeles Harbor and the mud flats of Alamitos Bay, San Gabriel River, Anaheim Bay and Newport Bay. All nereid species appear to be omnivorous; laboratory populations of nereids can feed and grow on a diet of algae, smaller crustaceans or various commercially prepared foods. The reproductive habits of nereids are varied. The most common method is to swarm to the surface, usually at night, release their gametes into the water where development occurs. The adults die after spawning. Other nereids, for example, *Neanthes arenaceodentata* do not swarm. Both sexes occupy the same tube where the eggs are laid and fertilized; the females dies but the male incubates the embryos for the next 3 weeks after which the young leave the tube and begin feeding. The male can reproduce again.

Family Nephtyidae (Figs. 53, 54). Two similarly appearing species of nephtyids are commonly encountered in Southern California. Both are flesh to gray in color; they can be distinguished by their habitat preference and by the greater amount of anterior pigmentation in *Nephtys caecoides* (Fig. 53). *Nephtys caecoides* prefers the finer sediments of the intertidal zone of Alamitos Bay, Anaheim Bay and Newport Bay; whereas, *Nephtys californiensis* (Fig. 54) is found in intertidal sandy beaches of Santa Monica, Long Beach, Huntington Beach and Newport Beach. Both species reach several inches in length. The food habits of nephtyids are poorly

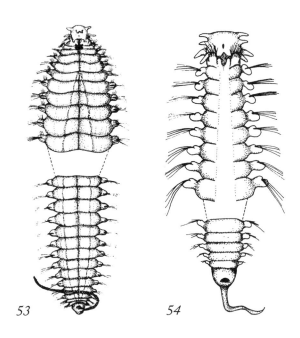

53   54

known; they are probably carnivorous since laboratory specimens will eat small polychaetes. They may also obtain nourishment from detritus present in the sediment.

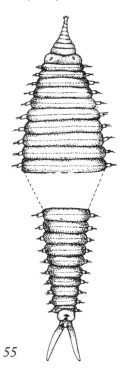

55

Family Glyceridae—the proboscis worms (Fig. 55). This family is characterized by having a long, tapered anterior end which terminates in front with 4 small tentacles. They possess a long, eversible proboscis which terminates in 4 black jaws; these jaws are capable of inflicting a wound when the worm is handled. Some people are allergic to the bite especially from the east coast species which is imported into Southern California for fish bait. Two similar appearing species are commonly encountered; both are flesh colored. *Glycera americana* (Fig. 55) may reach 15 inches in length; it burrows into the sand or mud in the intertidal and subtidal areas of outer Los Angeles-Long Beach Harbors, Alamitos Bay and Newport Bay. *Hemipodus borealis* measures up to 4 inches in length and has been taken from Cabrillo Beach and the sandy beaches of Long Beach and Newport Beach. It is probably a carnivorous feeder, but since it burrows in the sediment with its proboscis, it probably also feeds on detritus.

Family Goniadidae (Fig. 56). This family appears much like the glycerids by the possession

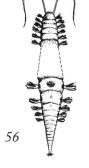

56

of a long prostomium. It is light brown in color with rows of darker spots along both the upper and lower surfaces. Specimens of *Goniada littorea* measure 3 inches in length and may be collected from the low intertidal zones in the muddy sand at Alamitos Bay, Anaheim Bay and Newport Bay.

Family Onuphidae (Fig. 57). The onuphids have 5 long tentacles with ringed bases arising from the head. They construct tubes which extend into the sediments. *Diopatra splendidissima* (Fig. 57) is an intertidal representative of onuphids. Its tube is composed of a parchment-like gray material which the animal secretes then places bits of shells, algae or other material to it as indicated in Fig. 57. Nothing is known of its food habits but it may be carnivorous as indicated by its large jaw apparatus. Fish or other animals nip off the anterior end of the worm since worms regenerating anterior ends are frequently seen. Specimens reach 8 inches with the tube extending more than twice this length. The tubes extend 1-2 inches above the surface which can be seen at low tide at Cabrillo Beach, Alamitos Bay, Anaheim Bay and Newport Bay. Tubes of this species or other onuphids are often washed up onto sandy beaches following rough seas.

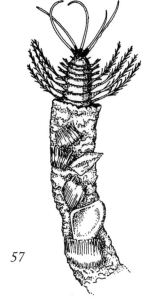

57

Family Eunididae (Fig. 58). The eunicids are similar to the onuphids; however, they may have 1, 3 or 5 tentacles on the prostomium which lack the ringed base. *Marphysa sanguinea* (Fig. 58) may reach 10 inches in length and is tan colored with red gills which arise from the parapodia; this species has been collected from Anaheim Bay and Newport Bay and Newport Bay from intertidal mud flats.

Family Lumbrineridae (Fig. 59). The lumbrinerids resemble earthworms in that their body is long, slender and the head lacks appendages. Their parapodia are weakly developed. Superficially all lumbrinerids resemble one another. *Lumbrineris erecta* (Fig. 59) is tan in color and specimens may reach 10 inches in length. It is generally found subtidally in muddy sands in Los Angeles-Long Beach Outer Harbors, Alamitos Bay and Newport Bay. An occasional specimen can be found intertidally in these areas. *Lumbrineris minima* is found with *Lumbrineris erecta*, but it only reaches 4 inches in length. *Lumbrineris zonata* is widely distributed in Southern California; specimens have been taken intertidally from rocky shores especially from the roots of the surf grass *Phyllospadix* (Fig. 572) and from sandy areas of Alamitos Bay and Newport Bay. It is deeply pigmented and often iridescent. Specimens measure up to 8 inches. Lumbrinerids take in sediment and digest the organic matter present much like an earthworm.

Family Arabellidae (Fig. 60). The arabellids resemble lumbrinerids; they differ in microscopic detail. *Arabella iricolor* (Fig. 60) can be recognized by the presence of 4 eyes on its

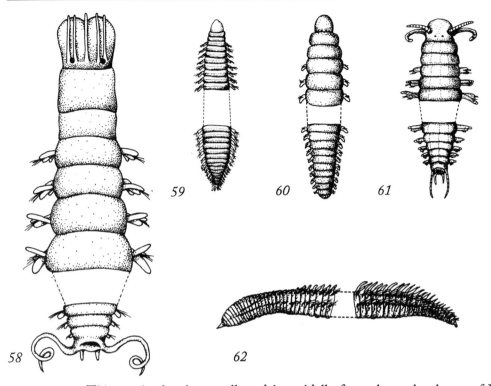

59 60 61 58 62

prostomium. This species has been collected intertidally from the rocky shores of Palos Verdes and Little Corona from sand and rubble under rocks and surf grass roots. They have similar feeding habits as the lumbrinerids.

Family Dorvilleidae (Fig. 61). The dorvilleids are small worms which bear 2 pairs of appendages on the head. *Schistomeringos longicornis* (Fig. 61) is a pale pink worm which measures up to 2 inches. This species has a behavioral characteristic which assists in identification; it rolls into a tight ball when disturbed. Large populations have been encountered from mussel beds on boat docks in Los Angeles Harbor; it also occurs in this niche in Marina del Rey, Alamitos Bay and Newport Bay. It occurs subtidally on the bottom of all bays, harbors and near many offshore domestic sewer outfalls. This species has been cultured in the laboratory where it fed on a diet of the green alga *Enteromorpha* sp. (Fig. 525). *Schistomeringos longicornis* has also been known under the names of *Dorvillea articulata* and *Stauronereis rudolphi*.

Family Orbiniidae (Fig. 62). The orbiniids may be distinguished from the other polychaete families by the presence of gills on the posterior end of the worm (Fig. 62, right side). The head lacks appendages. Two orbiniids are taken intertidally; *Leitoscoloplos elongatus* (Fig. 62) has a pointed head and *Naineris dendritica* has a rounded one. *Leitoscoloplos elongatus* is pink to tan in color, measures up to 10 inches and can be collected from Los Angeles-Long Beach Harbors, Alamitos Bay and Newport Bay in sandy mud both intertidally and subtidally. *Naineris dendritica* is more frequently encountered along the rocky shores in the roots of surf grass, but occasional specimens have been taken from the mussel clumps attached to pilings in Alamitos Bay and Newport Bay.

Family Spionidae (Figs. 63, 64). The spionids are a well represented family in Southern California waters. Superficially all spionids have a similar appearance (compare figures 63 and 64); it requires a specialist to distinguish the various species. All members build sand or mud tubes especially in quiet waters

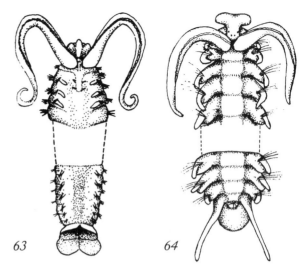

63   *64*

or bays. It is often possible to see countless numbers of small tubes (about 1/8 inch or less in diameter) projecting about 1/4-1/2 inch above the substrate in the intertidal zone. The anterior end bears a pair of long tentacles (palps) which it uses to filter material and food from the water. *Boccardia proboscidea* (Fig. 63) is darkly pigmented at its anterior end. It measures 1-2 inches in length and can be found abundantly along rocky shores in the upper zone tide pools wherever the rocks are of a softer nature such as found at Lunada Bay. The worm burrows into these softer rocks where large protuberances are found. It has also been collected in the intertidal sandy muds at the entrance of San Gabriel River. *Nerinides acuta* is a pale green spionid which forms sandy tubes in the mid-tide horizon of such sandy beaches as Santa Monica, Playa del Rey, Long Beach and Seal Beach. It possesses a pointed head. *Spiophanes missionensis* (Fig. 64) is one of the larger species of spionids; its tube is constructed of sand. It is tan to reddish-tan is color. It has been collected from the intertidal sand flats of Alamitos Bay, Newport Bay; subtidally at these localities and Marina del Rey, Los Angeles-Long Beach Harbors, and Anaheim Bay. Many of the spionids lay their eggs in capsules within their tube where early development takes. The young leave the capsule and become pelagic for a period of time. Generally spionid larvae are taken in plankton tows in inshore waters.

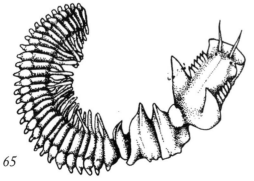

65

Family Chaetopteridae (Fig. 65). The chaetopterids are highly modified tube-dwelling polychaetes which possess 3 distinct body regions. *Chaetopteris variopedatus* (Fig. 65) constructs a U-shaped, parchment-like tube. The 2 opening may be 4-6 inches apart. The tube may be attached to the pilings as in Alamitos Bay or extend into the substrate at such localities as Lunada Bay and Portugese Bend. This worm is phosphorescent and has been the subject for many physiological experiments. *Chaetopterus* feeds by pumping water through its tube and capturing plankton on a mucus net it secretes; the worm then passes the net forward and is eaten.

Family Cirratulidae (Fig. 66). The cirratulids are characterized by a prostomium which lacks appendages; however, near the anterior end one or more pairs of thicker tentacles arise. In addition, a variable number of smaller paired gills arise from the lateral side of the worm. The gills may be limited to the anterior end or extend throughout most of the length of the worm. *Cirriformia luxuriosa* (Fig. 66) possesses many heavy, red tentacles near the anterior end and many pairs of red gills throughout most of the length of the worm. This is the largest of the local cirratulids; specimens may attain lengths of 6 inches. Specimens are more frequently encountered within protected waters; they have been taken from both floats and intertidal mud flats at Los Angeles Harbor, Alamitos Bay, Anaheim Bay and Newport Bay. They are present under rocks and in crevices at Lunada Bay, Flat Rock and Whites Pt. A complete life cycle takes 9 months to complete under laboratory conditions. Worms were fed the green alga *Enteromorpha*. A related species, *Cirriformia spirabrancha* is collected from fine sandy beaches.

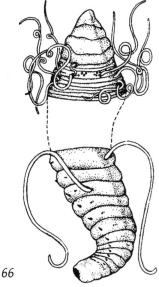

66

Family Opheliidae (Fig. 67). The opheliids are flesh-colored worms which are pointed at either end. *Armandia brevis* (Fig. 67) measures less than one inch in length. It has paired gills with dark eyespots between the gills in addition to the eyes on the head. Typically this

67

species is collected subtidally in all the bays and harbors but an occasional specimen can be taken from the bay mussel beds on boat floats from these same areas. A similar but smaller opheliid, *Polyophthalmus pictus*, also can occur with the mussel beds in bays and harbors but a more typical niche is within algal holdfasts at all rock jetties and rocky beaches in Southern California.

Family Capitellidae (Figs. 68, 69). The capitellids are generally reddish worms which resemble earthworms. *Capitella capitata* (Fig. 68) is abundant intertidally near the mouths of rivers such a Ballona Creek, Los Angeles River and San Gabriel River. It has been encountered in large

68

69

number in all bays and harbors if the water circulation is poor or if the area is polluted. This species also is taken in large numbers offshore near the vicinity of domestic outfall sewers. It is easily cultured in the laboratory, it will feed on a variety of foods. *Notomastus (Clistomastus) tenuis* (Fig. 69) may reach 10 inches in length. It occurs intertidally in lower Newport Bay as well as with surf grass roots at Flat Rock and Pt. Fermin. Capitellids engulf sediment for nourishment.

Family Maldanidae—the bamboo worms (Fig. 70). The maldanids are referred to as the bamboo worms because the widely separated segments (numbering 20 or less) give the appearance of joints. The maldanids are large worms which construct thick mud tubes. The anterior end lacks tentacles but varies in shape in the different species. The posterior end is also variously shaped and is frequently distinctive for the species. *Axiothella rubrocincta* (Fig. 70) measures from 4-6 inches. The posterior end is flower-like with alternately long and short petals (cirri) with one being longer than the rest. This species has

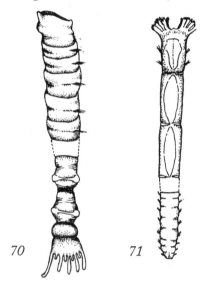

70   71

been taken intertidally from sandy mud at Lunada Bay, Alamitos Bay and Newport Bay.

Family Oweniidae (Fig. 71). The oweniids construct strong membranous tubes which are covered with coarse sand grains. *Owenia collaris* (Fig. 71) measures up to 4 inches in length; the head bears a circle of 6-8 branched tentacles. This gray colored worm has been collected intertidally from the sandy beaches of Long Beach; more extensive populations occur in subtidal waters.

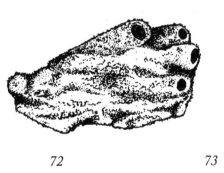

72

73

Family Sabellariidae (Figs. 72, 73). Members of this family construct heavy tubes which may occur singly or in massive clumps (Fig. 73). The worm is highly modified and possesses an operculum which closes the opening of the tube. *Phragmatopoma californica* (Figs. 72, 73), which is sometimes called the honey comb worm because of its tube construction, is the most commonly occurring polychaete in the rocky intertidal zone of Southern California. It has been observed at all rocky shores and rock jetties below the mussel beds. The honey-combed tube masses may occur with or beside the scaly worm snail *Serpulorbis squamigerus* (Fig. 136). The worm itself measures 2 inches in length and is purple in color.

Family Pectinariidae—the ice cream cone worm (Fig. 74). The pectinariids derive their common name from the shape of their tube (Fig. 74). The tube is constructed of brown sand grains which are only one sand grain in thickness; it is interesting to examine its tube under a dissecting scope. The worm secretes a mucoid material on which it places a selected sand grain. The mucus hardens to give the tube a definite shape. *Pectinaria californiensis* (Fig. 74) measures 2-3 inches. The anterior end contains many golden colored bristles in scoop-like arrangement. The larger sand particles are used for tube construction and the smaller ones eaten. Typically *Pectinaria californiensis* is collected subtidally from outer bays and harbors as well as off-shore waters; it has been taken intertidally during minus tides in Newport Bay.

74

Family Ampharetidae (Fig. 75). The ampharetids build tubes from mud or shell fragments. The worms are characterized by having many tentacles which retract into the mouth and a few pairs of non-retractile gills on the dorsal surface near the mouth. *Amphicteis scaphobranchiata* (Fig. 75) attains a length of 4 inches. This species has only been collected from the subtidal waters in Marina del Rey, Alamitos Bay and Newport Bay.

75

Family Terebellidae (Figs. 76-77). The terebellids are similar in appearance to the ampharetids. Their tubes are similar, but they differ in that the anterior tentacles do not retract into the mouth as in the ampharetids. The dorsal gills number from 1-3 pairs which may (Fig. 76) or may not (Fig. 77) be branched. *Pista alata* (Fig. 76) constructs a mud tube which may measure 4 inches. This species occurs subtidally in Alamitos Bay and Newport Bay. *Thelepus setosus* (Fig. 77) constructs a membrane tube covered with sand, shell fragments, bits of algae and other debris; it will measure up to 8 inches in length and 1/2 inch in diameter. An occasional specimen has

76

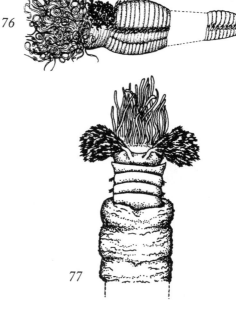

77

been collected from Pt. Fermin under rocks at low tide. All terebellid species feed on particulate matter; their ciliated tentacles may extend out 1 to 2 feet from their mouth to gather food.

Family Sabellidae—the feather dusters (Fig. 78). The feather dusters are so named because of the presence of numerous, pinnately-branched tentacles at the anterior end. The worms construct a mucoid tube to which sand grains adhere.*Megalomma pigmentum* (Fig. 78) measures 2 inches and has been taken from shallow waters in Alamitos Bay, Anaheim Bay, and Newport Bay. Another species, *Sabella media*, measures 4 inches in length and possesses a similar type of tube as *Megalomma pigmentum*. The gills of many sabellids, including *Megalomma*, bear eyespots which play a role in rapid withdrawl into its tube when a shadow is casted over the tube. The gills funtion *78* in respiration as well as a feeding mechanism.

Family Serpulidae—the calcareous tube worms (Figs. 79-85). The serpulids construct white tubes made of calcium carbonate (Figs. 79-81, 83-85). This material is secreted by the worms from cells located on the collar(the body region just posterior to the tentacles, see especially Fig. 84). The tubes may be long and intertwined with one another (Figs. 81, 83); they may be round in cross section (Figs. 79, 81, 83). They may possess a keel on the upper surface (Figs. 79, 85), or they may be minute in size and coiled (Fig. 84). The tentacles are feather-like as in the sabellid polychaetes. A funnel-shaped structure, the operculum, is present in most species (shown in Figs. 79, 80, 82, 85), and it functions as

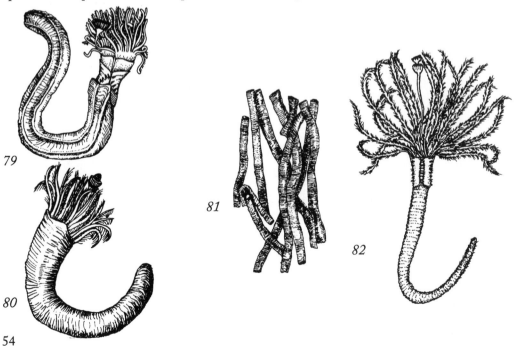

79

81

82

80

a lid which closes whenever the worm is disturbed. *Crucigera websteri* (Fig. 79) builds a tube having 3 longitudinal ridges on the upper surface. There are 4 spines at right angles at the base of the operculum (Fig. 79). This species occurs singly attached to shells or pilings in Alamitos Bay and Newport Bay. *Eupomatus gracilis* (Fig. 80) is similar to *Hydroides pacificus* (below). They differ in details of the outer circle of spines on the operculum; the margins of the spines are smooth in *Eupomatus gracilis* and lateral projects in *Hydroides pacificus* (requires a dissecting microscope to see). Tentacles are tan to

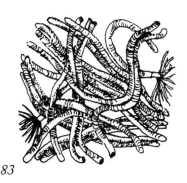

83

orange in color. This species attaches to boat floats in King's Harbor, Alamitos Bay, Newport Bay and to rocks at low tide at Pt. Fermin. *Hydroides pacificus* (Figs. 81, 82) may become very abundant on boat floats or on the sides of boats especially during the late summer and early fall when the water temperatures are warmer. Under these conditions, especially during years of El Niño, growth of the tube may be from 2-4 inches within a month. The tube masses generally fall during the first heavy rain of the winter. The tentacles may be variously colored; gray to tan tentacles are the most common color phase, but others may be purple, orange or red. The largest tube masses of *Hydroides pacificus* have been observed in Los Angeles-Long Beach Harbors; other localities where it has been seen are Marina del Rey, Alamitos Bay, Huntington Harbour and Newport Bay. *Salmacina tribranchiata* (Fig. 83) seldom reaches more than one inch in length but since it forms tube

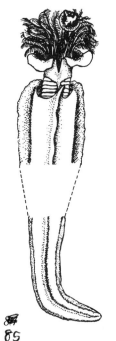

84
85

masses it can be seen under rocks along the Palos Verdes Peninsula and Little Corona. *Spirobranchus spinosus* (Fig. 85) is a serpulid having brilliantly colored tentacles. Red colored tentacles are the most common phase seen; but they can also occur in a red and white, orange or blue color phase. The tube is white with blotches of blue and has a high keel along the upper surface. Specimens are found attached to the undersurface of rocks at low tide at Lunada Bay, Pt. Fermin and Laguna Beach. At times they have settled on boat floats in Alamitos Bay. *Spirobranchus* is more common in shallow water at Santa Catalina Island than the mainland.

√ Spirorbids (Fig. 84) are often placed in Family Spirorbidae. These serpulids are small coiled shells which will measure 1/8 of an inch in diameter. There are many local species but generic and specific identifications are difficult. Two general groups are present-one in which the opening of the tube is to the right (Fig. 84) and

84

the other in which the tube opens to the left. These species can be very abundant in the rocky shore environment especially in the low tide zone where they attach to the under-surface of rocks. Less frequently they have been seen attached to the bay mussel in Alamitos Bay and Newport Bay.

The gills of serpulids function in the same manner as the sabellids.

PHYLUM ECHIURA
The echiuroids

The echiuroids are non-segmented, worm-like animals. The body is sausage-shaped with an anterior spoon-like proboscis. Because of the shape of the proboscis, the members of this phylum are often referred to as the spoon worms. Two heavy setae are present near the junction of the proboscis and body and a circle of setae around the anus in some species (Fig. 86). *Urechis caupo*-the innkeeper (Fig. 86). This flesh-colored species forms a U-shaped burrow in the intertidal and subtidal muddy sediments of Anaheim Bay and Newport Bay. Local specimens measure up to 6-8 inches. It feeds by pumping water through its burrow and trapping particulate matter on a mucous net which it later swallows. Many species of animals live within its burrow, such as polychaetes, clams, crabs and fish, which accounts for its common name.

Listrolobus pelodes (Fig. 87). These greenish colored worms are sausage shape and measure up to 4 inches in length. It has 2 stout setae located near the mouth. This worm lives in sediment in subtidal waters. *Listrolobus* will appear more-or-less suddenly in large numbers and disappear about 3 years later; all specimens are of the same age at a given locality. It is known off the Palos Verdes Pennisula and Santa Barbara-Ventura; specimens have been taken subtidally from outer Los Angeles—Long Beach Harbors.

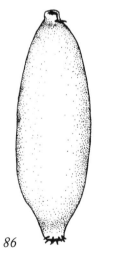
86

87

PHYLUM SIPUNCULA
The peanut worms

The sipunculoids (Figs. 88, 89) are worm-like animals and when in the contracted state they resemble peanuts. They possess a long stalk-like introvert which terminates with a circle of tentacles.

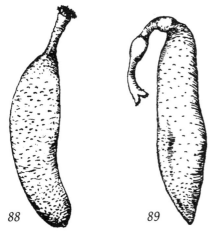

88   89

Thermiste pyroides, known also as *Dendrostomum pyroides*, (Fig. 88). This species is pale to brown in color and is characterized by having 4 branched tentacles around its mouth. Some black thorn-like structures are present on the body. Local specimens have been observed in the rubble under rocks and in abandoned burrows of rock-boring clams all along the Palos Verdes Peninsula and Little Corona. *Phascolosoma agassizi* (Fig. 89). Specimens are generally brown in color with black pigmented spots. The introvert ends with a circle of many unbranched tentacles. Many rings of hooks are present just posterior to the mouth. Specimens from Southern California are 1-2 inches long; they have been collected at all rocky shores especially from the roots of surf grass or in sand under rocks in the intertidal zone.

PHYLUM MOLLUSCA
The molluscs

The molluscs are one of the largest groups of animals in the marine environment in terms of number of species; they include 5 Classes: Polyplacophora, or chitons, Gastropoda, or snails and sea slugs, Bivalvia, or bivalves (clams), Scaphopoda, or tusk shells, and Cephalopoda, or octopus and squid. All but the scaphopods are represented intertidally; these can be dredged from shallow, off shore waters. The majority of species of molluscs are characterized by some sort of a shell or shells which is secreted by the mantle. The shell may be external, as in the chitons (Fig. 90), snails (Fig. 95) and clams (Figs. 201, 202); or it may be internal, as in the squids; or it may be secondarily lost, as in the sea slugs. An additional characteristic feature is the possession of a strong muscular foot which is used in locomotion. Probably 300-400 species of molluscs occur intertidally in Southern California shores, about 150 of which are described herein.

CLASS POLYPLACOPHORA (formerly Class Amphineura)—the chitons (Figs. 90 94).
The chitons have a rounded upper surface and a flattened lower surface. The upper surface has 8 calcareous plates which fit into one another. Separately these plates are referred to as butterfly shells because of their shape. The plates are variously ornamented in the different species which is used in specific identification. A girdle surrounds the calcareous plates. The girdle may bear granules, scales or hairs on its surface depending upon the species. The lower surface of the chiton consists of a central large, oval shaped muscular foot which is separated from the margin by a groove known as the mantle cavity. Branched gills may be seen segmentally arranged in this cavity. Chitons are herbivores; they feed on algae.

Family Lepidochitonidae (Figs. 90, 91). The girdle is characterized by spines or spicules but not hairy. The valves are variously sculptured.

Cyanoplax hartwegii (Fig. 90). This species of chiton measures up to 2 inches in length. The surface of the girdle is granulated with some wart-like granules. The olive green to brown valves often have white streaks along the posterior edge. It is found associated with *Pelvetia fastigiata* (Fig. 546) which constitutes its principle diet. It has been found in the mid-tide zone at the rocky shores at Palos Verdes and Little Corona.

90

Nuttalina fluxa (Fig. 91). This species of chiton is known by the common name of rough chiton, troglodyte chiton and California chiton (the former scientific name was *Nuttalina californica*). It is identified by its small size of 1.5 inches or less in length, the girdle is alternately light and dark and possessing spines and shells of brown, pink and white colors. Coralline algae is the principle food for this species. Sea gulls are known to feed upon the California chiton. *Nuttalina fluxa* is the most frequently encountered chiton in Southern California; it lives in pits in the rocks in the upper and mid-tide horizon at all rocky shores and many rock jetties in Southern California.

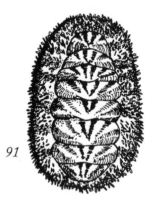

91

Family Mopaliidae (Fig. 92). The girdle is characterized with either setae (long hairs) or with spicules.

Mopalia mucosa—**the mossy chiton** (Fig. 92). This species measures up to 2.5 inches in length. The 8 shells are generally dull gray or black in color and have some fine longitudinal lines on the surface. The girdle bears dark colored stiff curled hairs hence the common name. The appearance of the chiton may be altered as the result of other organisms, such as seaweeds, worms or moss animals growing on the shells and girdle. The mossy chiton maintains a home from which it emerges at night to feed upon algae. It returns to its home after traveling up to 2 feet in distance. The mossy chiton is common in Southern California marine habitats. In the rocky environment it is found in the mid tide zone generally within rock crevices. It has been collected from the boat floats of the principal bays and harbors in the area.

Family Ischnochitonidae (Figs. 93, 94). The girdle is scaley and the valves scultured in various patterns.

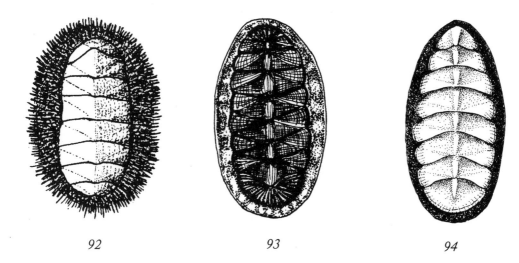

92   93   94

Lepidozona pectinulata (Fig. 93).   This chiton is oval in shape and large specimens are over 1.5 inches in length. The girdle is covered with small overlapping scales. The valves and girdle are greenish or brownish yellow in color. This species is commonly found under rocks at low tide at all rocky shores of Southern California.

Stenoplax conspicua—the conspicuous chiton (Fig. 94).   This is the largest species of chiton encountered in Southern California; specimens measure up to 4 inches long. The shells are pale green on the sides and the mid regions are generally pink in color as a result of the erosion of the outer green layer of shell. The outer surface of the girdle is composed of large, overlapping scales. This species feeds upon different species of algae, and it is known to be fed upon by octopus. The conspicuous chiton can be collected at all Southern California rocky shores from the undersurface of rocks present in the low tide zone.

CLASS GASTROPODA—the snails and sea slugs (Figs. 95-199).   The gastropoda, or stomach footed animals, are characterized by the possession of a single shell. The shell maybe coiled (Figs. 119-126), twisted (Fig. 136), uncoiled (Figs. 99-117), internal (Fig. 190), or secondarily lost (Figs. 191-199). The shell is secreted by the mantle which is the soft tissue located just within the shell. The shell of snails consists of an opening, the aperature, and one or more whorls which are separated from one another by grooves or sutures. The shape and form of shells vary considerably from species to species. When disturbed, the marine snail withdraws into its shell and closes the aperature with an operculum.

Subclass Prosobranchiata—Species with shells in which the gills and mantle cavity are situated anteriorly over the head. This group includes most marine shelled gastropods.

Order Archaeogastropoda—Primitive gastropods characterized by having two gills, two auricles to the heart and two kidneys. This group is primarily marine shelled gastropods.

Family Haliotidae—the abalones (Fig. 97). The abalones are characterized by large slightly coiled, ear shaped shells which have a series of open and secondarily closed holes. The number of holes varies with the species and these openings function as exits for water which passes over their gills. The inside of the shell is pearly in appearance. The larvae of the red abalone (adults not figured) settle on crustose coralline algae. It has been demonstrated the settlement and metamorphosis is induced by the chemical g-aminobutyric acid. The crustose coralline algae contain a red accesory photosynthic pigment, phycoerythrobilin, which is chemically related to g-aminobutyric acid. This acid has been utilized in inducing larval metamorphosis in commercial culturing of abalone. Commercial culturing of the red abalone for human consumption occurs north of Los Angeles County. It is necessary to feed abalone the giant kelp *Macrocystis pyrifera* (Fig. 540) which is collected from the ocean. Abalone are harvested when they reach about 3 inches in length. Many species occur in Southern California, most of which are subtidal. Abalones are protected by law. Each species has a minimum legal size and maximum number which can be taken in a day. Consult the nearest California Department of Fish and Game office for the license required, the season, the size and number which can be collected.

95

96

Haliotis cracherodii—the black abalone (Fig. 95). This species of abalone rarely gets 5 inches in length; it has a smooth shell which is dark green to black. There are usually 5 to 7 open holes in the shell. The black abalone feeds on drift algae especially the larger brown species such as *Macrocystis pyrifera* (Fig. 540) or *Egregia laevigata* (Fig. 542). Enemies include octopus, starfish, fish and man. This species is the most frequently seen intertidal abalone. Small specimens can be seen on the undersurface of rocks in the intertidal zone. It has been observed at all the rocky shores in Los Angeles and Orange Counties.

Haliotus corrugata—the pink abalone (Fig. 96). This species of abalone may reach 10 inches in length. The shell is pink and the surface rough. Two to 4 holes remain open. The surface of the shell is generally covered with fouling organisms such as serpulid polychaetes (Figs. 80, 83, 85), scaley worm snail (Fig. 136), bryozoans and small

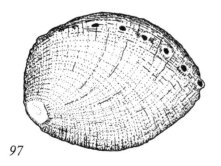

97

algae. It feeds on drift brown and red algae. It is rarely seen in the intertidally, but occurs subtidally in the rocy areas of Palos Verdes Peninsula.

Haliotus fulgens—**the green abalone** (Fig. 97). This species of abalone is olive green to reddish brown in color and reaches up to 8 inches in length. It has 5 to 7 open holes. The surface is sometimes overgrown with fouling organisms as are the pink abalone. It feeds on drift brown and red algae. It is generally found in subtidal areas where the surf is strong. While not as common as in the past, it is seen in the Palos Verdes Peninsula rocky areas.

Family Fissurellidae—the key hole limpets (Figs. 98 100). The common name of this family has its origin from the presence of an opening at the apex of the shell. This hole serves as an exit for water which enters the mantle cavity and passes over the gills where the oxygen carbon dioxide exchange takes place.

Fissurella volcano—**the volcano limpet** (Fig. 98). An oval aperature is present at the apex of the shell. Fine radiating ridges extend out from the apex to the margin. The shell is pink in color with darker colors present as blotches. Specimens are generally one inch or less in length. The volcano limpet is a commonly encountered species at all rocky shores and rocky jetties of Southern California especially under rocks in the mid to low tide horizon.

98

Megathura crenulata—**giant key hole limpet** (Figs. 99-100). A large species which attains a shell length of 5 inches and an overall body length of up to 8 inches. The black colored mantle extends over the shell. The yellow colored foot is located on the under surface of the animal. It feeds on algae and colonial tunicates. The giant key-hole limpet has been used as food and the Indians used the shells for wampum. The giant key hole limpet has been collected from the low intertidal zone among rocks along the

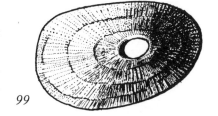

99

100

Palos Verdes Peninsula, Dana Pt. and along the jetty of Alamitos Bay.

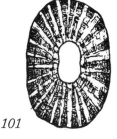

Megatebennus bimaculatus—**the two-spotted key-hole limpet** (Fig. 101). The shell length is less than an inch and the overall lenth including the soft parts is an inch. The shell is gray in color with bands radiating out from the hole. The opening of the shell is hour glass in shape. It has been collected intertidally from under rocks or attached to the holdfasts of algae, but it is more common subtidally from the rocky shores of Southern California.

101

Family Acmaeidae—the limpets (Figs. 102-117). The limpet or China hat shells are so named because of their resemblance to the hat of Chinese laborers. The position of the apex, or top of the shell, varies with the species, but it is always located at the anterior half of the animal. The color and markings on the undersurface of limpets are useful in identification of the different species (Figs. 104, 106, 108, for example). A muscle attaches to the shell; this muscle leaves a scar on the shell and, in the case of the owl limpet (Fig. 117), the scar forms part of the outline of an owl. A limpet feeds upon algae which it scrapes from the surface of rocks with its rasp like teeth or radula. Many of the species with the generic name of *Collisella* are also known under the former generic name *Acmaea*. The generic name *Acmaea* is still valid for some of the species of limpets.

102

Acmaea mitra—the white cap limpet (Fig. 102). The shells measure 1.5 inches in length and nearly that in height. The shells are white inside and out, but it is often covered with coralline algae. It feeds largely on coralline algae. It is found in the lower intertidal to subtidal waters attached to rocks covered with coralline algae. Empty shells are often washed up on rocky shores such as Little Corona and the Palos Verdes Peninsula.

Collisella pelta—the shield limpet (Figs. 103, 104). The apex is near the anterior end. Ribs of various strength radiate out from the apex. The shell is heavier than most species and inflated, giving it convex sides. The outer color is gray or white with the ribs darker; the interior is white with a blue cast and the margins are alternately light and dark. Specimens will on occasion measure greater than one inch. The shield limpet feeds upon erect species of brown and red algae. This species has been collected from the mid to low tide level at such rocky shores as Pt. Dume, Lunada Bay, Flat Rock, White's Pt., Pt. Fermin, Little Corona, Laguna Beach and Dana Pt. An occasional specimen was taken from the boat floats at Huntington Harbour and Newport Bay.

Collisella digitalis—fingered limpet (Figs. 105, 106). The apex of the shell is near the anterior end (Fig. 105) which frequently forms a concave margin. Ribs radiate away from the apex giving the shell an undulating margin (Fig. 106). The interior of the shell has alternating light and dark margins and a brownish stain near the center. The finger limpet is one of the characteristic species of the high intertidal rocky shores (Fig. 10). The fingered limpet feeds upon microscopic algae which grows on the rock surfaces especially in crevices or shaded areas. Specimens up to one inch are found clustered together especially within crevices at all rocky shores and jetties of Los Angeles-Orange Counties.

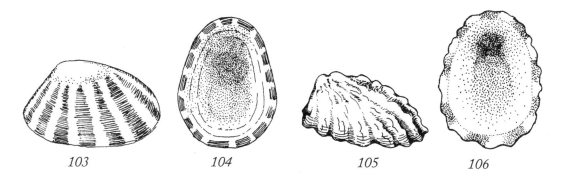

103   *104*   *105*   *106*

Collisella scabra—**the rough or ribbed limpet** (Figs. 107, 108).   This species derives its two common names from the heavy radiating ribs which give the surface a rough appearance. The margin of the shell is irregular in shape because of these radiating ribs (Fig. 107). The undersurface is primarily white, but it may have a darker margin and a darkened line where the muscle attaches. The shell measures one inch or less in length. *Collisella scabra* often occurs with *Collisella digitalis*, and they both feed on microscopic algae growing on the surface of the rocks. It has been collected from the mid tide horizon at all rocky shores visited and from the rock jetties at Alamitos Bay and Newport Bay.

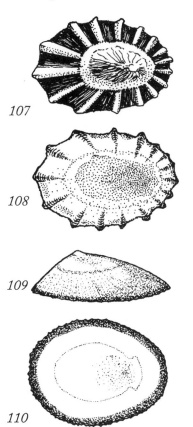

107

108

✓*Collisella limatula*—**the file limpet** (Figs. 109, 110). The upper surface has many radiating rows of small file teeth, hence its common name. The surface is dark brown in color which may be checkered with light brown spots. The file limpet may be distinguished from the other limpets by its dark margin on the undersurface against a white background (Fig. 110). The shell measures up to 1.5 inches. The file limpet moves when covered with water to feed on microscopic algae. It is found at many localities and habitats in southern California. It is abundant in the mid-to lower tidal zones at all rocky shores; it is also present on boat floats in all bays and harbors of Southern California.

109

Collisella strigatella (Figs. 111, 112). This species of limpet measures up to 0.75 inches in length. The apex is located at the anterior third of the shell. The outer shell is olive to blue in color and is often eroded. The interior of the shell is bluish white with brown

110

111   112   113

markrings in the center. It is found at the rocky shores of Southern California in the mid-tidal zone.

Collisella asmi—the **black limpet** (Fig. 113). This small limpet measures less than 0.5 inches in length. It is found on the shells of the snails *Tegula funebralis* (Fig. 125) and *Tegula gallina* (Fig. 127). It feeds upon the microscopic algae growing on the surface of the snail. It is found at all rocky shores in tide pools where these two species of *Tegula* are found.

114

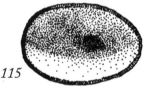

115

Notoacmaea incessa—the **seaweed or kelp limpet** (Figs. 114, 115)). This small limpet, which measures 0.75 inches or less, is so named because it lives on the stem of brown algae. The sides are nearly parallel (Fig. 114) and if placed on a flat surface, the shell will rock. The color is dark brown both inside and out with a darker colored margin on the inside (Fig. 115). This species is found in the low tide horizon wherever brown algae, such as the feathery boa *Egregia laevigata* (Fig. 542), is found attached to rocks. The limpet feeds upon the kelp which can lead to the destruction of the sea weed.

Lottia gigantea—the **owl limpet** (Figs. 116, 117). The common name comes from the outline of the muscle attachment scar and from the scar from the tissue surrounding the mantle cavity on the undersurface of the shell (Fig. 117). This species is the largest one of local limpets; specimens may reach up to 3 inches in length. The owl limpet is active when covered with water; it may move several inches while feeding during high tide. Field studies have indicated that *Lottia gigantea* will frequently return "home" after feeding during a high tide cycle.

116   117

While its primary food is microscopic algae, it will feed on newly settled barnacles or sea anemones in order to maintain "its territory." This species has been collected from the mid- to upper tide zone at all rocky shores in Southern California. Specimens have also been taken near the entrance of Newport Bay on rocks and in Alamitos Bay along the cemented sea wall in the marinas.

Family Trochidae—the top shells (Figs. 118-128). These shells are so named because of their resemblance to a child's toy top. The interior shell margin is pearly. This is a common family of snails in Southern California waters.

Calliostoma tricolor—**the three colored top shell** (Fig. 118). The shell is so named because of its yellow brown background color with alternating rings of brown and white. The shell will obtain heights of nearly one inch. It has been collected from the low tide zone at Pt. Fermin and Little Corona.

118

Norrisia norrisi—**Norris' top shell** (Figs. 119, 120). An easily identified shell which is colored brownish red and pearly white interiorly. It has an opened umbilicus (Fig. 120). Shells are generally one inch in height but can reach 2 inches. Living specimens are found on brown algae, especially *Egregia laevigata* (Fig. 542) and *Macrocystis pyrifera* (Fig. 540),in the low tide horizon at Palos Verdes Peninsula, Little Corona and Laguna Beach. Empty shells are frequently seen in rubble along these shores.

Tegula aureotincta—**the gilded turban or tegula snail** (Figs. 121, 122). One of the 4 species of this genus commonly found in Southern California. The gilded turban can be distin- guished from the others by its open umbilicus (Fig. 122) which is yellow orange in color. The shell is light in color varying from green to gray to brown. Shell height reaches 1.5 inches. This species has been collected at low tides from many of the rocky shores of the Palos Verdes Peninsula and Little Corona.

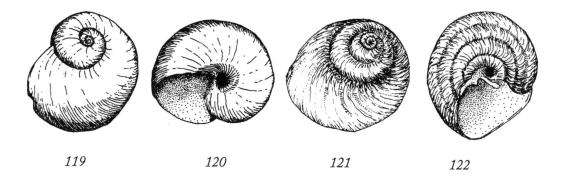

119   *120*   *121*   *122*

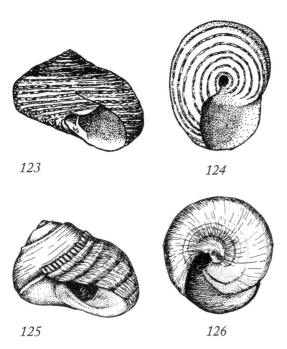

123   *124*

125   *126*

Tegula eiseni—the banded turban (Figs. 123, 124). This species, known also as *Tegula ligulata*, has numerous, evenly shaped ridges which may take on a beaded appearance. The shell is brown and will reach one inch in height. It has an open umbilicus (Fig. 124). This species has been collected at all rocky shores from the low tide horizon.

Tegula funebralis—the black turban (Figs. 125, 126). This snail is black in color with a purple tinge. The apex of larger specimens generally erodes away giving it a pearly white appearance. The aperture is white with a black border along the outer lip. This species lacks an umbilicus (Fig. 126). Shells measure one inch in height. This species feeds on microscopic red and green algae. The slipper limpet *Crepidula perforans* (Figs. 144, 145) lives just within the aperature of the black turban. A very common species in the mid tide horizon and in tide pools at all rocky shores and on rock jetties.

Tegula gallina—the speckled turban (Figs. 127, 128). As the common name suggests this snail can be identified by its speckled white, green or gray spots on a dark background. This species lacks an umbilicus (Fig. 128). The shell height is generally about one inch. It is another common top shell, and it is frequently found associated with *Tegula funebralis* (Fig. 125) at the local rocky shores.

Family Turbinidae also known as the top shells (Fig. 129). This family is similar to the Trochidae; they differ in that the Turbinidae have calcareous opercula instead of membranous ones.

Astraea undosa—the wavy top turban (Fig. 129). An easily identified snail by its large size up to 4 or 5 inches in height, and its wavy ridges. The operculum is calcareous with heavy, outer ridges. Small sized specimens can be seen at minus tides at any of the rocky shores in Southern California especially associated with brown algae.

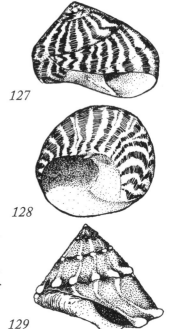

127

128

129

ORDER MESOGASTROPODA—gastropod snails characterized by having only one gill, one auricle to the heart and one kidney.

Family Littorinidae—The periwinkles (Figs. 130-133). Small snails with only 3 4 whorls which are a characteristic animal of the high tide and splash zone on rocky shores (Figs. 10). Both species of littorines feed on algal films composed of diatoms, blue green algae and green algae.

Littorina planaxis—the gray littorine (Figs. 130, 131). The shell is gray in color except for a white line which can be seen along the lower margin of the aperture. This species, as in the case of most littorines, is gregarious in nature. Large specimens will reach 0.5 inches in height. Groups of them may be observed in rock crevices in the high tide horizon or splash zone at any rocky shore in Southern California. Specimens also occur on the rock jetties at the entrances to bays and harbors.

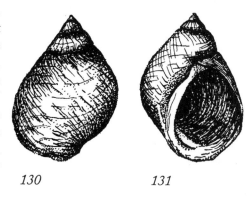

130   *131*

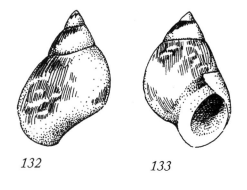

132   *133*

Littorina scutulata—the checkered periwinkle (Figs. 132, 133). Not as common as the gray littorine; it can be distinguished from it by the light spots on the outer surface of the shell and the lack of the white line in the aperture. Specimens may reach 0.5 inches in height. The checkered periwinkle occurs in the high tide horizon, rarely as high as the splash zone. It has been observed at the rocky shores off the Palos Verdes Peninsula, Little Corona and Laguna Beach.

Family Caecidae (Fig. 134). Small tubular shells with an operculum at one end which becomes spiraled.

Caecum californica (Fig. 134). A light brown, tubular shell with 30-40 closely spaced annulations. This shell measures about 0.25 inches in length. It is present in sand and gravel in the low intertidal tide pools especially wherever eel or surf grass is found.

134

Family Turritellidae (Fig. 135). Shells elongated with many whorls.

Turritella cooperi—Cooper's turret (Fig. 135). A long, narrow shell with over 10 body whorls. Two prominent ridges occur in each whorl with some lesser ones as well. It is light buff in color. Cooper's turret is a subtidal species which lives in sand or fine sediments. Dead specimens are frequently found washed up on the shores of Southern California.

Family Vermitidae—the worm snails (Fig. 136). These snails construct long, winding, uncoiled tubes which resemble polychaete worm tubes. The tubes are attached to rocks or animals.

135

Serpulorbis squamigerus—**the scaly worm shell** (Fig. 136). This species is one of the characteristic animals of the mid tide horizon of rocky shores. The tube masses, which may measure several inches in thickness, are attached to large rocks. The tubes measure up to 0.25 inches in diameter and up to one foot in length. The circular ridges give the appearance of scales which accounts for its common name. One should be especially careful where the scaly worm shell lives since the shell at the opening is extremely sharp and can easily cut one's finger. It feeds by secreting mucus which traps diatoms and small animals. It can exposed this mucus net for 30 minutes after which it takes it into its mouth. The young develop in a capsule near the entrance of the tube and are released as young larvae. *Serpulorbis squamigerus* is common at all rocky shores in Southern California; it is also found on the rock jetties at the entrances of the bays and harbors.

136

Family Potamididae—the horn shells (Fig. 137). Multi-spiraled snail which is brownish in color; an anterior canal present which is weakly developed.

137

Cerithidea californica—**the California horn snail** (Fig. 137). This is one of the more commonly encountered snails in bays. It is found by the thousands in the inner reaches of a bay on mud flats in the high tide zone (Fig. 13). The mature shell is black in color, measures one inch in length and possesses 9-10 body whorls. It crawls about on the mud flats and feeds by scooping up organic debris and microscopic algae with its radula. The California horn snail is the intermediate host for many marine parasitic trematodes; the adult stage occurs in shore birds. The largest populations of the California horn snail occur in Anaheim Bay and Upper Newport Bay. It can also be collected in Ballona Creek, Cabrillo Beach in Los Angeles Harbor and Alamitos Bay.

Family Epitoniidae—the wentletraps (Figs. 138, 139). Small, beautiful white shells with well developed ridges and rounded aperatures. The word wentlewrap means winding staircase which these species resemble.

Epitonium tinctum—the tinted wentlewraps (Fig. 138). A small white shell with about 10 whorls and measures slightly less than 1 inch. A characteristic brown line occurs in the suture. (The separation groove between two whorls.) Heavy ribs (or varices), numbering about 12, extend from one whorl to the next. This species feeds on the tentacles of sea animals under laboratory conditions. Specimens have been collected at low tides in the rubble under rocks along the Palos Verdes Peninsula, Little Corona and Laguna Beach.

138   139

Opalia funiculata—sculptured wentletrap (Fig. 139). Shells are less than 1 inch in height. It is white to golden yellow in color. The ribs are weakly developed and number about 14 per body whorl. It feeds on sea anemones. It is found at the bases of sea anemones, especially *Anthopleura xanthogrammica* (Fig. 36), in the low intertidal rocky shores of Southern California.

Family Hipponicidae—hoof shells (Fig. 140). Limpet-shaped animals in which there are circular ridges; the muscle scar is horse-shoe shaped.

140

Hipponix tumens—ribbed hoof snail (Fig. 140). Shell is white and attains lengths of 0.5 inches. The apex is anterior and curved downward which gives it an unique appearance in side view (Fig. 140). The shell is rough textured with concentric ridges. The ribbed hoof shell is found at rocky shores in rock crevices at low tides.

Family Calyptraeidae—the slipper limpets (Figs. 141-149). This family is characterized by having a shell plate within the opening giving it an appearance of a slipper. The muscles of the animal attach to this plate.

Crepidula aculeata—spiny slipper shell (Fig. 141). This whitish shell has a shelf which is notched at each side. It has spines on spiral ridges which accounts for its common name. It reaches 0.5 inches in length. It is found subtidally at rocky shores; specimens wash up on the shores of rocky beaches.

Crepidula onyx—the onyx slipper shell (Figs. 142, 143). The name onyx is derived from the claw like appearance of the curved apex of the shell. The apex is located at the anterior end of the shell (the end with the plate inside). The shape of the onyx slipper shell is extremely variable. The shell is dark brown in color except for the whitish shell plate. Large shells reach 1.5 inches in length. This species is gregarious and many specimens have been seen on top of one another. The larger specimens on the bottom are females, and the small

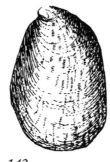

141   142   143

ones on top are males; those in the middle are undergoing a sex change. It feeds on organic debris and plankton. This species is especially common on rocks, pilings and floats in Marina del Rey, Alamitos Bay, Anaheim Bay, Huntington Harbour and to a lesser extent Los Angeles Long Beach Harbors. Some have been collected from the rocky intertidal shores at White's Pt. and Little Corona.

Crepidula perforans—**western white slipper shell** (Figs. 144, 145). It is usually a small shell, but specimens greater than one inch in length are known. The shape of the shell is rectangular and curved. It is found in a variety of niches at rocky shores including attached to rocks, in the aperature of empty gastropod shells occupied by hermit crabs and empty burrows in rocks.

144   145

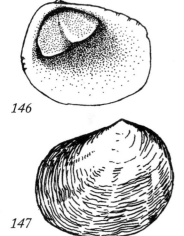

146

147

Crepipatella lingulata—**the half slipper shell** (Figs. 146, 147). This slipper shell can be distinguished by its mottled brown and white color and by the detachment of the shelf on the left side of the shell. *Crepipatella lingulata* is found at low tides attached to rocks at Lunada Bay, Flatrock, White's Pt., Newport Bay jetty and Little Corona.

Crucibulum spinosum—**the spiny cup and saucer shell** (Figs. 148, 149). This limpet shaped shell has many curved spines on the outer surface and a funnel shaped shell attached to the inside of the apex. These two characteristics, which enables identification readily, account for its common name spiny cup and saucer shell. The shell is generally white to tan and can be 1 inch in height. As characteristic for the members of this family, *C. spinosum* is a filter feeder. Living specimens are not commonly encountered in Southern California, but empty shells are common in intertidal shell rubble. Specimens have been collected at low tide on rocks near the entrance of Newport Bay.

148   149

Family Naticidae—the moon snails

(Figs. 150-153). These snails are large, have a heavy shell, and are globular in shape. The operculum is made of horny material.

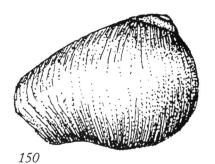

150

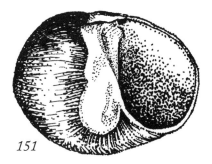

151

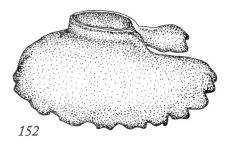

152

Poliaices

Neverita reclusianus—**the southern moon snail** (Figs. 150-152). This large snail reaches 2-3 inches in diameter. The shell is tan to brown in color, very thick and heavy. The eggs are laid in a collar consisting of cemented sand grains; the collar will measure up to 6 inches in diameter (Fig. 152). They are called sand collars and may be found along intertidal sandy beaches. This species is carnivorous in habit and feeds upon bivalves; it drills a hole in the shell near the umbo. Specimens have been found at low tides on either side of the peninsula separating Alamitos Bay and the ocean in Long Beach and in Newport Bay. Living specimens of *Neverita reclusianus* and the larger, northern moon snail species *Polinices lewisii* are under protection by the California Department of Fish and Game. These species were formerly known under the genus *Polinices.*

Family Triviidae (Figs. 153, 154). Cowry-shaped shells with an elongated aperature; the outer lip is folded in and both the outer and inner lips have ridges.

Trivia solandri—**large coffee bean shell** (Figs. 153, 154). The common name of this animal stems from its brown color, its size of about 0.5-0.75 inches in length and its coffee bean like appearance. The teeth extend from the aperture, around each side to near the whitish line

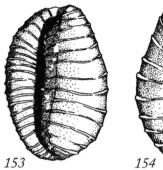

153   154

along the mid region of the upper surface (Fig. 154). This species is not seen too often in Southern California, partially because of its small size; it has been collected at White's Pt. and Laguna Beach during minus tides in association with algae. It crawls over the surface of compound ascidians and eats the individual zooids as well as the common tunic. This species is sometimes referred to as *Pusula solandri*.

Family Cyraeidae—The cowries (Figs. 155, 156). The cowries are among some the most beautiful shells in the world. The shell is highly polished and is protected from wear by the mantle tissue. The aperture is elongated, infolded and both margins provided with fine teeth.

Cypraea spadicea—the chestnut brown cowry (Figs. 155, 156). This is the only true cowry present in Southern California waters. Specimens will reach 2.5 inches in length. The surface near the aperture and the sides are white to tan in color with the upper surface chestnut brown. This species is carnivorous feeding on such invertebrates as sponges, sea anemones, tunicates, etc. Living specimens are not often seen in the intertidal zone, probably as a result of the tendency of most people to keep any specimen encountered. Specimens have been seen under rocks and on rock ledges in the low tidal zone at Pt. Dume, Lunada Bay, White's Pt., Pt. Fermin and Little Corona. This species has also been referred to as *Zonaria spadicea*; *Zonaria* is a subgenus of the genus *Cypraea*.

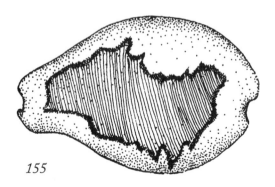
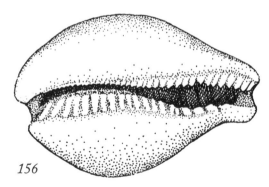

155   156

Order Neogastropoda—These snails are characterized by a long siphon for water intake and they often have a canal (see Figs. 159, 162) at the lower part of the shell which protects the siphon. They are largely carnivores and feed on clams as well as other invertebrates.

Family Muricidae—the rock shells (Figs. 157-166). The rock shells are rough appearing shells with regularly spaced ridges (varices).

Ceratostoma nuttalli—Nuttall's hornmouth shell (Figs. 157, 158). This shell measures 2 inches in length. Its color varies from white to dark brown. Three longitudinal curved ridges extend the full length of the shell. One large, spinous tooth is found at the lower edge of the aperture on the outer lip. The common name hornmouth comes from the presence of this tooth (or horn) near the mouth. A closed canal leads from the aperture to the lower end of the shell. This snail is common at low tides along the Palos Verdes Peninsula, on pilings in Alamitos Bay and in Little Corona. It is a carnivorous species feeding on mussels and other clams. This species is known also under the scientific names of *Pterynotus nuttalli* and *Purpura nuttalli*.

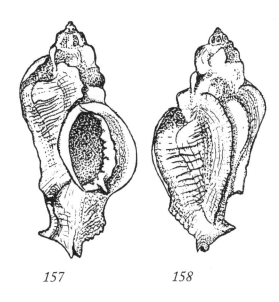

157   158

159

Pteropurpura festivus—festive murex or festive rock shell (Fig. 159). This species of snail may be distinguished by its curved ridges and fine brown lines running around the shell. The aperture is nearly circular in outline and the canal closed. It has been collected from among rocks at minus tides along the Palos Verdes Peninsula, Little Corona, Laguna Beach and within Newport Bay. This species is known also by the scientific names *Shaskyus festivus* and *Jaton festivus*.

Maxwellia gemma—gem murex or gem rock shell (Figs. 160, 161). An easily identified snail by the broad black circular stripes over a white shell. The aperture is circular and canal closed. The shell reaches 1 inch in height. The gem murex lives at rocky shores at low to mid tide levels; it has been collected at many of the rocky shores along the Palos Verdes Peninsula and Little Corona.

Ocenebra circumtexta—the circled rock shell (Figs. 162, 163). This species of

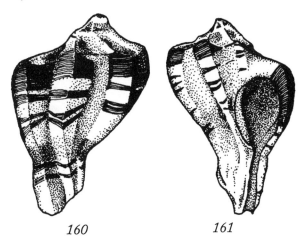

160   161

73

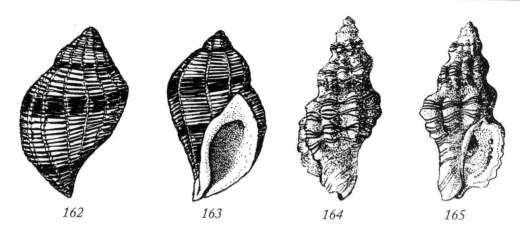

162   163   164   165

Ocenebra can be distinguished by the numerous, regular, spiral ridges to each body whorl. The shell is whitish with dark circular bands of brown. The shell attains lengths of about 1 inch. It has been found intertidally at low tide at Lunada Bay, White's Pt., Pt. Fermin and Laguna Beach.

Roperia poulsoni—**Poulson's rock shell** (Figs. 164, 165). This snail can be separated from *Ocenebra circumtexta* by its dark gray color with many fine brown lines encircling the whorls. Many knob like teeth are located within the outer lip. A narrow, open canal leads from the aperture. Specimens will measure over 1 inch in length. This snail feeds on both species of *Mytilus*. The rock crab *Cancer antennarius* (Fig. 338) feeds upon Poulson's rock shell. This species is especially common on the rock jetties leading into Marina del Rey, Alamitos Bay and Newport Bay. Specimens are occasionally taken from rocks in the mid tide horizon at Lunada Bay, White's Pt., Pt. Fermin, Little Corona and Laguna Beach. This species is also known as *Ocenebra poulsoni*.

Forreria belcheri—**Belcher's murex** (Fig. 166). A large snail measuring up to 5-6 inches in height. It has many large spinose ridges along the body whorls which were formed during the resting stage in growth. The aperture is extended into a long, open canal. It is found in subtidal sandy bottoms and sometimes within bays or harbors. It is more commonly seen as empty shells washed up on beaches.

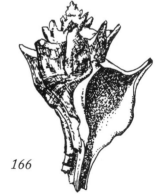

166

Family Thaididae (Figs. 167-170). These snails possess a large aperture and body whorls without varices.

Acanthina spirata—**the angular unicorn shell** (Figs. 167, 168). Shell measure about 1 inch in length and have 5-6 body whorls. Each whorl has many light colored spiral ridges and dark colored spiral lines. An open canal leads from the aperture to the lower end of the shell. A single spine extends out from the lower margin of the outer lip. Many small teeth are present in a line within the outer lip. The angular unicorn shell is carnivorous and probably feeds upon littorine snails. This species has been collected in abundance from the intertidal zone along the rock jetties leading into Alamitos Bay and Newport Bay. Specimens are also found on the rocks in the low tide zone at Flatrock, Lunada Bay, White's Pt., Pt. Fermin, Little Corona and Laguna Beach.

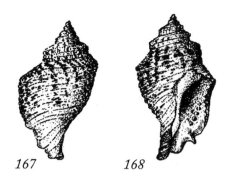

167   *168*

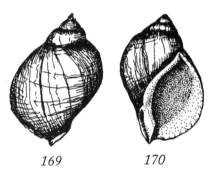

169   *170*

Nucella emarginata—**emarginate dogwinkle or rock thais** (Figs. 169, 170). The aperture extends more than half the length of the shell. A short, open canal leads down from the aperature. The shell has about 3 whorls, dark brown in color and specimens will reach 1.5 inches in height. This snail is carnivorous and feeds upon *Mytilus californianus* (Figs. 201, 202). It has been found on rocks in the intertidal zone especially within mussel beds at the King's Harbor jetty, Pt. Fermin, Little Corona and Laguna Beach. This species is also known under the name *Thais emerginata*.

Family Buccinidae (Fig. 171). Heavy shells with an elongated, open canal.

Kelletia kelletii—**Kellet's whelk** (Fig. 171). A large shell reaching up to 7 inches in height. It is whitish in color with many darkened, spiral groves. An open canal extends down from the aperature. It is a subtidal species occurring on rocky bottoms in Southern California.

Family Collumbellidae—**the dove shells** (Figs. 172, 173). Small snails which have a broad notch at the end of an open canal.

Amphissa versicolor—**the variegated amphissa or variegated dove shell.** (Figs. 172, 173). The shell is mottled with mixtures of gray, brown, and white. It has raised ridges running both directions on a whorl. Specimens measure about 0.5-0.75 inches in height. It has been collected from the undersurface of rocks in the low tide zone at Lunada Bay, White's Pt. and Pt. Fermin.

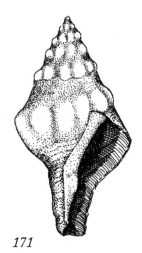

171

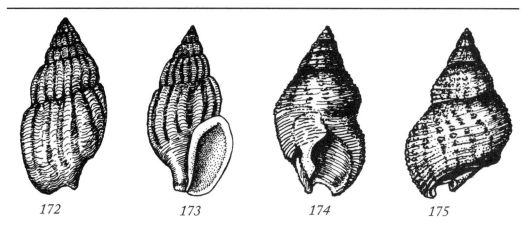

172   *173*   *174*   *175*

Family Nassariidae—basket shells (Figs. 174-177). Shells of this family are referred to as basket shells because of the sculpturing of the surface which resembles a woven basket. Most of the species in this family feed upon bivalves or dead animals.

Nassarius fossatus—**channeled basket shell** (Figs. 174, 175). This species is so named because of the characteristic channel, or groove which extends around the base of the shell (Fig. 175). The channel basket shell may reach 2 inches in height; it is colored yellow-gray outside and orange within the aperture. This species of basket shell feeds on dead animals. This species lives on the intertidal sandy mud flats of Alamitos Bay and Newport Bay.

Nassarius mendicus—**the lean basket shell** (Figs. 176, 177). This species of *Nassarius* is narrow in relation to its height. The longitudinal ridges are often prominent. Height is generally 0.5-0.75 inches. It is light brown in color with lighter colored aperture. It has been collected at low tide from the rocks along the Palos Verdes Peninsula and Laguna Beach.

Family Mitridae—the miter shells (Figs. 178, 179). This family has shells which are narrow and pointed. The open canal is short.

Mitra idae—**Ida's miter** (Figs. 178, 179). This snail is easily identified by its thick, black covering (periostracum), its shape and dark color. Three ridges (or folds) are present along the inner margin of the aperture (Fig. 179). Shells reach 1 inch in length and are found under rocks along the Palos Verdes Peninsula and Laguna Beach.

Family Olividae—the olive shells (Figs. 180, 181). The olive shells are highly polished, oval-shaped shells.

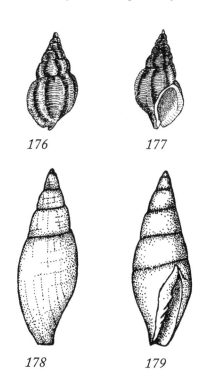

176   *177*

178   *179*

76

Olivella baetica—little olive shell or beatic olivella (Fig. 180). This species is similar but smaller (0.5 inches) than the more common California purple olive shell (Fig. 181). The shell is polished and grayish brown in color. It is found in shallow subtidal waters in the bays and harbors of Southern California.

Olivella biplicata—the California purple olive shell (Fig. 181). The last body whorl is large and overlaps the aperture (Fig. 150). Specimens are generally 0.5-0.75 inches in length. The color is variable, ranging from nearly white to a dark gray or brown

180

with a tinge of purple. These shells were strung and used as ornaments or money by the Indians of Southern California. The purple olive shell feeds on decaying plants and animals. Enemies of this species include the octopus, *Conus californiicus* (Fig. 182), *Neverita reclusianus* (Fig. 150), *Pisaster brevispinus* and *Astropecten armatus* (Fig. 379). *Olivella biplicata* lives just beneath the sand in the low tide zones of the sandy beaches of Southern Califonia. Empty shells are frequently found not only here but also at rocky shores.

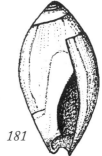

181

Family Conidae—the cone shells (Fig. 182). Cone-shaped shells with an aperture which extends most of the length of the shell. The radula(teeth) has been modified to a harpoon-shaped structure which injects poison into its prey. Some species of cone shells in other parts of the world are extremely toxic to humans, but the local species is not; however, care should be exercised in handling living specimens of the California cone shell. The poison is a protein which paralysis the prey. It then takes in the entire animal into its digestive tract.

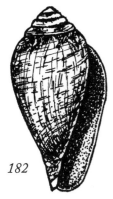

Conus californicus—the California cone shell (Fig. 182). A small brown-colored cone shell which measures less than 1 inch in length. It feeds on such species of snails and clams as *Olivella biplicata* (Fig. 181), *Nassarius* spp., *Bulla gouldiana* (Fig. 185), *Macoma nasuta* (Fig. 242), *Tagelus californianus* (Fig. 257) and many species of polychaetes. The Indians used the shells as money and in some of their religious activities. Empty shells are often found in the rubble under rocks at rocky shores. Living specimens of the California cone shell can be found in this environment at all rocky shores in Los Angeles and Orange Counties.

Subclass Opistobranchiata—the sea slugs and sea hares. This group is entirely marine in habitat. They either have reduced shells, internal shells or lack them entirely.

182

Order Tectibranchiata (Figs. 183-189). External (Figs. 183-187, 189) or internal (188) shell present; one gill present.

Family Acteonidae (Fig. 183). Small shells with the body whorls finely sculptured; aperature elongated.

Rictaxis punctocaelatus—striped barrel snail (Figs. 183, 184). It is a white shell with two broad bands of spiral pitted lines. The aperature is long and narrow extending the length of the final body whorl. The striped barrel snail reaches 0.5 inches in height. It is common in such areas as upper Newport Bay and Alamitos bay as well as other protected waters of Southern California. It may be seen at low tides, it is more frequently present in shallow subtidal waters. This species is also known under the generic name *Acteon*.

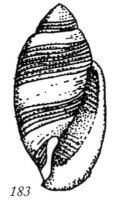

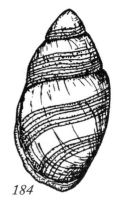

183   184

Family Bullidae—the bubble snails (Figs. 185, 186). The shell of this family is very thin, the apex is sunken, and the outer lip greatly expanded. This family is separated from the closely related Family Atyidae (below) by not having the outer lip flared at the lower margin (compare Fig. 186 with 187).

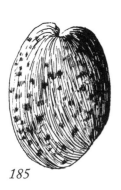

185   186

Bulla gouldiana—the cloudy bubble snail (Figs. 185, 186). The color of the shell varies; it has a background color of gray to brown with darker spots scattered over the surface. The soft parts are yellow-orange in color. The shell often reaches 2 inches in length. Egg strings of *Bulla gouldiana* are often seen on mud flats during the summer months. They are pale yellow in color and appear much like spagetti. The bubble snails is a herbivore and it may also feed upon small snails and bivalves. *Bulla gouldiana* has been collected in the mud flats at Cabrillo Beach in Los Angeles Harbor, Alamitos Bay, Anaheim Bay and Newport Bay.

Family Atyidae—the paper bubble shells (Fig. 187). Characterized by having a flared, outer lip as discussed above.

Haminoea virescens—the green paper bubble snail (Fig. 187). The thin shell is pale green with some yellow pigment present. It is smaller than the cloudy bubble snail; specimens may reach 0.75 inches in length. *Haminoea virescens* is a rocky shore inhabitant; it has been found at White's Pt. and Pt. Fermin at low tides.

187

Family Scaphandridiae (Fig. 188).   Shells cylindrical in shape with an elongated aperture which extends much of the length of the shell.

Acteocina inculata (Fig. 188).   A small, light brown shell measuring about 0.3 inches in height. The aperture is long and narrow extending the length of the last body whorl. It is found intertidally on the mud flats of upper Newport Bay and other such environments.

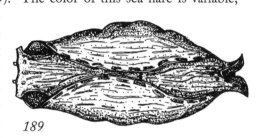

188

Family Aglajidae (Fig. 189).   These species lack rhinophores which are tentacle-like structures over the head.

Navanax inermis—**the striped sea hare** (Fig. 189).   The color of this sea hare is variable; different combinations of white, yellow, blue and brown occur. Specimens may obtain lengths of 7 inches. *Navanax inermis* is carnivorous; it feeds upon bubble snails or it is even cannibalistic on occasion. Whitish eggs are laid in long, thread-like gelatinous strings. The species lacks a shell. The striped sea hare has been found on the sandy mud flats of Alamitos Bay and Newport Bay at low tide.

189

Family Aplysiidae—**the sea hare** (Fig. 190).   A pair of rhinophores, or tentacle-like structures located over the head, is present giving the animal a hare-like appearance.

Aplysia californica—**the sea hare** (Fig. 190).   This species of seahare may attain lengths of 15 inches and weigh over 15 pounds. Typically it is purple in color, but specimens of reddish-brown to dark green may be encountered. The yellowish eggs are laid in strings and deposited as gelatinous entangled masses on rocks or sea weeds. When the sea hare is handled, it gives off a purple fluid which is annoying but harmless to humans. The purple color comes from the red pigment present in red algae which is a principle food item in its diet. The unique nature of its nervous system makes the neurons ideal for studying

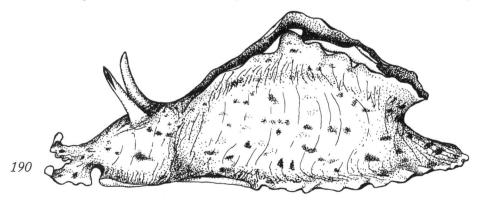

190

neurological responses. *Aplysia californica* is more frequently found intertidally at all rocky shores in Los Angeles and Orange Counties and less frequently on the sandy mud flats or Alamitos Bay and Newport Bay.

Order Nudibranchiata—the sea slugs or nudibranchs (Figs. 191-199). Considered by many to be the most beautiful of the marine animals. They are small and often inconspicuous animals in the marine environment. Their bright and varied colors make them prize specimens in marine aquaria. The shell and true gills are absent, but newly, secondary, acquired gills may be present as retractile groups on the upper surface at the posterior end (Figs. 191-195) or as small finger-like processes all along the upper surface (Figs. 197-199). Each species lays eggs in a characteristic mass and frequently in marine aquaria. Family characteristics are based on both external and internal structures, some of which are difficult for the non-specialist to see; therefore, families are not included for the nudibranchs.

Archidoris montereyensis—**the dorid sea slug** (Fig. 191). This species reaches 2 inches in length. The basic color is pale yellow-brown with brown patches scattered over the upper surface. The branchial plume is whitish. Eggs are laid in white, coiled ribbons. Sponges are an important item in the diet of this nudibranch. This species has been collected from the rocky shores of White's Pt., Pt. Fermin, Laguna Beach and along the jetty of Alamitos Bay.

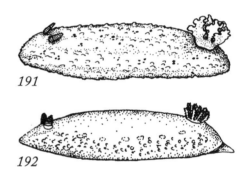

Cadlina flavomaculata (Fig. 192). This species of sea slug measures up to 1 inch in length. It is pale yellow to cream colored with paired rows of yellow-brown spots on the upper surface. The rhinophores (or tentacles) are dark brown and the gills white. It has been observed feeding on sponges. Specimens are typically found in tide pools along rocky shores such as those at Lunada Bay, Flatrock Cove and White's Pt.

Diaulula sandiegensis (Fig. 193). This nudibranch is easily distinguished from the other species in Southern California by the presence of several darkly colored rings on the upper surface. The background color is pale yellow to brown. Specimens measure 2-3 inches in length. The eggs are white and deposited as coiled ribbons. It feeds on sponges. *Diaulula sandiegensis* has been collected from most of the rocky shores of Palos Verdes Peninsula and the jetty at Alamitos Bay.

Rostanga pulchra—**the red sponge sea slug** (Fig. 194). A small (about 0.5 inch), red colored nudibrach which is frequently found on the red sponge, *Plocamia karykma* (Fig. 19). The white rhinophores are surrounded by a sheath which is white at its base and red at its end. The gills are white. This species of nudibranch has been collected at low tide at the rocky shores of Flatrock Cove, Pt. Fermin and Laguna Beach.

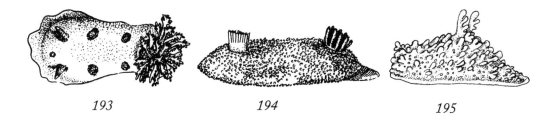

193   194   195

Aegires albopunctatus (Fig. 195). This pale yellow sea slug is generally 0.5 inches in length. Brown spots are scattered about the upper surface which has many tentacles. The gills are white in color and 3 in number. Eggs are deposited in short, narrow, white spiral bands. Specimens are found in protected niches at low tide at all rocky localities visited along the Palos Verdes Peninsula.

Polycera atra (Fig. 196). This species has a background color of pale gray to dark gray, almost black, with longitudinal lines of darker pig-ment. Orange spots are scattered about the upper surface. The gills number up to 11 finely branched structures which are light in color with orange tips. This species is readily identified by its color and by the pointed processes which extend forward from the head. Specimens measure 0.5-0.75 inches in length. It feeds on ectoprocts such as *Bugula* spp.

196

(Figs. 360, 362) and *Membranopora* spp. (Fig. 374). It opens up the individual zooid of the colonial ectoproct and sucks the soft parts out. *Polycera atra* has been taken frequently from the boat floats in many localities on Los Angeles-Long Beach Harbors; it has also been collected at Pt. Fermin and Alamitos Bay.

Coryphella trilineata (Fig. 197). This nudibranch measures about 1 inch in length. The body is pale gray in color with a white line extending down the mid-dorsal line. The cerata are brown in color. The rhinophores have spiral lamellae throughout most of its length. It feeds on hydroids such as *Tubu-laria crocea* (Fig. 27). The eggs are deposited in spi-ral loops. Specimens have been taken from different hydroid species in Alami-tos Bay.

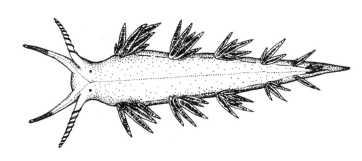

197

198   199

Capellinia rustya (Fig. 198). The animal is pale in color which ranges from a gray to a yellow to green. The cerata appear brownish because of the liver tissue within. Flecks of darker pigment are scattered over the body. Specimens range in size from about 0.25 to 0.4 inches in length. It has been taken at low tide from hydroid colonies at Flatrock Cove.

Hermissenda crassicornis (Fig. 199). This beautiful nudibranch measures up to 2 inches in length. The colors are variable and bright with various combinations of green, blue, orange, yellow, white, purple and brown. The cerata are translucent in appearance and are variously colored on the surface; the internal brown color is from the pigment of the liver within. It has a varied diet including hydroids, ectoprocts, small sea anemones and colonial tunicates. *Hermissenda crassicornis* is common in Southern California and it does well in marine aquaria. It is generally found wherever the hydroids are found. The eggs are white to pink in color and are laid in a spiral ribbon. This nudibranch has been found in all the local bays and harbors crawling on boat floats and pilings. Offshore it has been collected from Flatrock Cove, White's Pt. and Pt. Fermin.

CLASS BIVALIA—The bivalves or pelecypods (Figs. 200-272). The pelecypods, or hatchet-footed animals, are characterized by having 2 shells which fit over the soft parts of the animal. Both shells are usually alike but in some cases, especially in those species that have one shell attached to rocks, the 2 shells are different (Figs. 210, 217, 218). The shells are secreted by the mantle which is the tissue located within the edge of the shell and extends around the margin except at the point of attachment of the 2 shells. The 2 shells are held together by a secreted ligament and are often held in position by teeth located inside (Figs. 234, 236, 243). The right and left shells of bivalves may be determined in one of two ways: (1) hold the 2 shells in both hands and point the beaks (the narrowed end in Figs. 200-204) and thumbs in the same direction; the right shell will then be in your right hand, etc. or (2) if the ligament (point of attachment of the 2 shells) is held up and towards your body, then the right shell will be in your right hand, etc. The different species of bivalves differ externally in their shape, color, and sculpturing of the shell and internally by the number and structure of the teeth, if present, and by the scars from muscle and mantle tissue attachment points.

Family Mytilidae—the mussels (Figs. 200-207). The shells are dark brown to deep purple or black in color and equal in size. They attach themselves to solid objects with thread-like secretions called byssal threads. These threads are secreted by a byssal gland; this secretion flows down a groove in the extended foot to a solid object; the foot moves away when the secretion is hardened. The local mussels are used primarily as fish bait; they were used earlier as food by the coastal Indians.

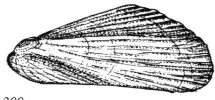

200

Geukensia demissus—the ribbed horsemussel (Fig. 200). It is also known under the generic names of *Modiolus* and *Ischadium*. This species was accidentally introduced into San Francisco Bay during the latter part of the past century from the Atlantic coast, presumably by way of oyster shipments. Later it appeared in Newport Bay, Anaheim Bay and Alamitos Bay by unknown means. The shell measures up to 4 inches in length and is dark brown in color. The shell has many radiating ribs especially at the posterior end of the shell. The clapper rail (Fig. 499) is known to feed on the ribbed horse mussel. It is found in the upper reaches of these bays in the upper tidal zones attached to rocks or present within burrows made by the yellow shore crab *Hemigrapsus oregonensis* (Fig. 343).

Mytilus californianus—the California mussel (Figs. 201, 202). The shell is covered with a black, thin periostracum, which is often worn away exposing a dark purple to white shell. The California mussel is a conspicuous animal along the rocky shores of California where specimens may reach 10 inches in length. It is a convenient indicator organism for the mid-tide horizon (Fig. 10). The pea crab *Fabia subquadrata* (Fig. 350) is sometimes found living within the mantle cavity of the mussel. The California mussel has many enemies including several species of snails, starfish and sea otters. The mussels was an important component of the diet of the coastal Indians, and many people today collect the mussels for food and fish bait. Specimens are present at all rocky shores, on offshore pilings (Fig. 12), and on rock jetties leading into bays and harbors.

✓ *Mytilus edulis*—the bay mussel (Figs. 203, 204). This species is smaller, smoother and more wedge-shaped than the California mussel. The periostracum is generally not eroded away. Specimens of the bay mussel measure up to 3-4 inches in length and are blue-black in color. The bay mussel is the most conspicuous organism in bays and harbors of Southern California. It is the dominant organism found attached to boat floats, pilings (Fig. 11) and sea walls. The bay mussel has similar enemies as the California mussel plus such crabs as

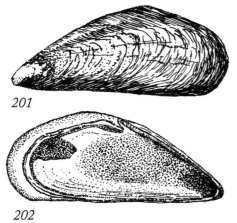

201

202

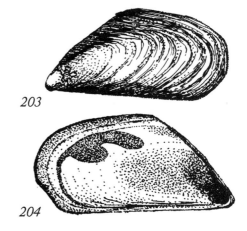

203

204

Pachygrapus crassipes (Fig. 345) and *Cancer antennarius* (Fig. 338). It may be seen occasionally on offshore pilings and attached to rocks and in crevices at rocky shores such as White's Pt. and Pt. Fermin.

There has been some question that the Pacific Coast population of *Mytilus edulis* is not the same species that occurs in northern Europe as determined by electrophoretic evidence (biochemical determination). It is believed that the population in Southern California is similar to *Mytilus galloprovincialis* from the Mediterranean Sea. The population in Central and Northern California is similar to *Mytilus trossulus*, and the population in Oregon northward is considered to be *Mytilus edulis*. However, since these species of mussels appear similar to the observer, the name *Mytilus edulis* is used in this book.

205

206

207

Septifer bifurcatus—branched ribbed mussel (Fig. 205). A small mussel that will measure 1.5 inches in length. It is black in color with many fine, closely spaced ribs extending the lenth of the mussel. The posterior margin in crenulated. It is found intertidally under rocks or sometimes among *Mytilus californianus* (Figs. 201, 202).

Modiolus rectus—fat horse mussel or straight horse mussel (Fig. 206). A large, thin shelled mussel which measures up to 5 inches in length. The surface of the shell is smooth and dark brown in color. It lives singly unlike most mussels on mud flats especially in Seal Beach Naval Weapons Station and upper Newport Bay.

Lithophaga plumula—date mussel (Fig. 207). Burrows into rocks aided by secretion of an acid and it lines its burrow with a calcareous deposit. It attains a length of about 1.5 inches. It attaches itself to the inside of the burrow with byssal threads and is found in the intertidal zone at rocky shores.

Family Ostreidae—the oysters (Figs. 208, 209). This family includes the edible oysters. The shells vary greatly in shape, form and structure. The surface of the shells have many irregular ridges or flutes. The shells are usually white or gray in color and may appear pearly especially inside. The left shell attaches to solid objects such as rocks or pilings.

Ostrea lurida—the native oyster (Figs. 208, 209). The shells are white to gray in color and are provided with many thin, scalloped flutes. The shells are irregular in shape which is the result of conforming to the contours of the substrate or other animals. Shells generally

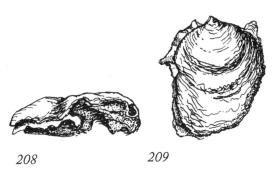

208　　　　209

reach 2-2.5 inches in length. This species is a commercially important species from Morro Bay northward. It is cultivated in some bays north of Point Conception, however the Japanese oyster *Crassostrea gigas* is cultivated more extensively because of its larger size. It is common but never abundant on boat floats, pilings and rocks in Marina del Rey, Alamitos Bay, Anaheim Bay, Huntington Harbour and Newport Bay.

Family Pectinidae—the scallops, pectens or fan shells (Figs. 210-213). The common names are derived from the use of the muscle which holds the 2 shells together as food (scallops) and from the shape of the shell. Many ribs radiate out from the area of the hinge; the number of these ribs may be of assistance in identifying some species of pectens. Two equal (Fig. 213) or unequal (Fig. 212) projections, termed ears, extend out from the area of the hinge. The majority of the pectens swim in a backwards direction by flapping their 2 shells together.

Crassedoma gigantea—**the purple-hinged pecten** (Figs. 210-211). This species may be easily identified by the purple color on the inside of each shell near the area of the hinge. The purple hinged pecten may attain lengths of 6 inches. After a free-swimming stage this species attaches its right shell to rocks or pilings. The purple-hinged pecten has been reported to live as much as 25 years. The muscle which holds the two shells together is eaten and tastes much like the commercial scallops. It has been collected from the jetties,

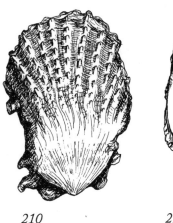

210　　　　　　　211

pilings and floats in Alamitos Bay and Newport Bay and from the undersurface of rocks at low tide at Little Corona and Laguna Beach. The purple hinged pecten has also been known under the names *Hinnites multirugosus* and *Hinnites giganteus.*

212

Leptopecten latiaratus—**the broad-eared pecten** (Fig. 212). This small pecten measures only 1 to1.5 inches in diameter. It has 12-15 ribs radiating out from the hinge. The ears are large, flattened and unequal in size. The shells are yellow to brown in color with zigzag lines of white. It attaches to rocks, pilings, eel grass and other objects.

This species has only been taken intertidally attached to boat floats in Alamitos Bay and Newport Bay. It is typically found offshore attached to kelp.

Argopecten aequisulcatus—the **speckled scallop** (Fig. 213). The beautiful, symmetrical pecten measures up to 3 inches in length. It has 19-22 flattened ribs which radiate out from the hinge. The ears are symmetrical. The shell is yellow-orange to reddish-brown in color. It will attach to rocks or other structures, but it is not permanent. Specimens have been observed on the bottoms of Alamitos Bay and Newport Bay at low tide. The species is protected by law at all times. Empty shells are often seen washed up on sandy beaches. This species is also known as *Plagioctenium circularis*.

Family Limidae—the file shells (Fig. 214). Light colored shells with each shell asymmetrical and having poorly developed ears. They are capable of swimming much the same way as pectens.

Lima hemphilli—the **file shell** (Fig. 214). White shells with faint radiating ribs. Specimens are about 1 inch in length. Easily identified by their numerous tentacles which extend out from the shells which the shells can not completely enclosed. This species is typically found offshore in subtidal waters; it has been found often among the bay mussels nestled among the mussels and other organims attached to boat floats in Marina del Rey. This species is known also as *Lima dehiscens*.

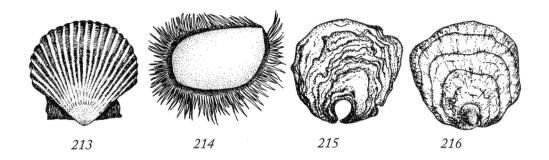

213   *214*   *215*   *216*

Family Anomiidae—the **jingle shells** (Figs. 215, 216). This family is known by this common name because the shells are so thin that they jingle when dropped. The right or lower shell attaches to solid objects by way of a byssal plug which extends through a circular opening near the hinge.

Anomia perviviana—the **pearly jingle** (Figs. 215, 216). A thin shell with a pearly appearance on the inside hence its common name. Color of outer shell varies but is usually an off white. The lower shell attaches to a rock or other attached animals with the shape varying according to where it is found. It measures up to 2 inches in diameter and is found attached to clams (often) or rocks off shore in Southern California. The upper shell often washes ashore.

Pododesmus cepio—the abalone jingle (Fig. 217). The common name comes from its occasional attachment to the upper surface of abalones. The shells measure about 2-3 inches in diameter. It is gray white in color and occasionally iridescent. The left shell (Fig. 217) has radiating ribs around the margin. Many animals attach to the top shell including boring sponges, sessile polychaetes and pholad clams. Specimens have been collected intertidally in Orange County from the jetty at Newport Bay and from the rocks at low tide at Laguna Beach.

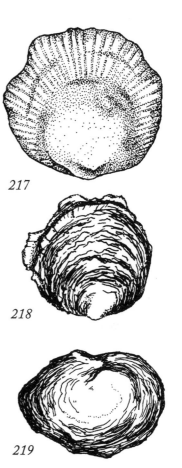

217

Family Chamidae—the rock oysters (Figs. 218, 219). The members of this family are always attached to solid objects. The shells are unequal in size and shape as a result of the contours of the rock and are roughly sculptured with frills.

218

Chama arcana (Fig. 218). The shells are white to translucent in uneroded specimens. The left shell attaches to the rock or piling. The curve of the upper, or right shell, is to the right when the shell is held upward (Fig. 218). Large specimens may reach 2 inches in diameter. It has been commonly found attached to pilings in Alamitos Bay and Newport Bay. Occasional specimens have been collected at low tide attached to rocks at White's Pt., Pt. Fermin and Little Corona. This species has been referred as *Chama pellucida*, a species known only from South America.

219

Pseudochama exogyra—the **California reversed chama** (Fig. 219). This species is referred to as reversed chama because the right shell is attached to the rocks and the curve (counter-clockwise) of the upper, or left shell, is to the left in contrast to *Chama arcana*. The shells are generally white and reach 2 inches in diameter. This species has been collected from the mid-tide zone from pilings and rock jetties in Alamitos Bay and Newport Bay.

Family Carditidae (Fig. 220). The beaks, or point of attachment of the two shells, near the anterior end; hinge with cardinal teeth; outer surface of shell with prominent ribs.

Glans subquadrata—**little heart clam** (Fig. 220). A heavy, small bivalve measuring less than 0.5 inches. It has about 15 strong ribs radiating out from the point

220

of attachment which gives a crenulating margin. The little heart clam is found at rocky shores under rocks at low tide. This species is known also under the name *Glans carpenteri*.

Family Lucinidae (Fig. 221).   Circular shaped shells with the beaks in the center; cardinal teeth present; external ligament.

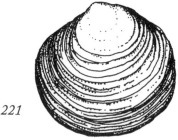

Epilucina californica—**California lucine** (Fig. 221).   A white shell measuring about 1.5 inches in diameter. It has many fine concentric ridges. This species is present in gravel at rocky shores at low and subtidal depths in Southern California.

221

Family Cardiidae—**the cockles or heart clams** (Figs. 222-224).   The word cockle is derived from the Latin word meaning shell; the common name heart shells, stems from the outline which resembles a heart when the 2 shells are viewed from the side. Ribs radiate out from the area of the hinge.

Laevicardium substriatum—**the egg-shell cockle** (Figs. 222, 223).   The shells are oval, inflated, thin and the ribs faint. The shell is tan with reddish-brown spots, especially on the inside of the shell. Specimens reach 1 inch in length. This species has only been taken from subtidal water in Alamitos Bay; however, undoubtedly it occurs in the other bays.

Trachycardium quadragenarium—**spiny cockle** (Fig. 224).   The common name is derived from the presence of spines present on the heavy radiating ribs. The margins of the shell are crenulated. The spiny cockle is light brown in color and measures 3 inches in diameter. It is found in upper Newport Bay as well as in off shore sandy bottoms.

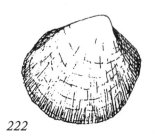

222   *223*   *224*

Family Veneridae—**the venus clams** (Figs. 225-237)   Large conspicuous hinge teeth are present on the inside of the shells. The outside of the shells are often sculptured with ribs which radiate out from the area of the hinge and concentric growth ridges which may be raised. The common name of the family comes from the genus *Venus* of which the shells of some species resemble the marble used in the famous statue.

Chione californiensis—**the banded chione, banded cockle or California chione** (Figs. 225, 226).   This species may be distinguished from the other two species of the genus *Chione*

 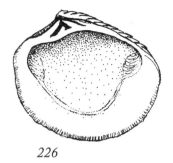

225   226

(see below) by possession of heavy, widely-spaced, concentric growth ridges (Fig. 225). The largest specimens reach 2.5 inches in length. The shells are white in color. This species lives just below the surface of intertidal or subtidal sandy muds. It has been collected from Alamitos Bay, Anaheim Bay and Newport Bay.

Chione fluctifraga—**the smooth chione or smooth cockle** (Figs. 227, 228). The ribs and concentric growth ridges are weakly developed which gives the outer surface of the shell a smooth appearance. A pallial sinus occurs in this species but not in the other 2 species of *Chione* (compare the faint lines in Fig. 229 with Figs. 226, 230). The outside of the shell is often chalky white and the inside is white with purple pigmentation along the border by the muscle scars. It measures as much as 3.5 inches in length. People collect this species of chione and use it making a chowder. The smooth chione lives just beneath the substrate both intertidally and subtidally in the back bay areas of Alamitos Bay, Anaheim Bay and Newport Bay.

227   228

Chione undatella—**the wavy chione or wavy cockle** (Figs. 229, 230). The concentric growth ridges are wavy, prominent and set close to one another. The ribs are fairly noticeable over the shell. This species measures up to 2.5 inches in length and the shell is white on both surfaces. This species is also used in making a chowder. *Chione undatella* has been collected intertidally from just beneath sandy muds from Alamitos Bay and Newport Bay. Occasional specimens occur intertidally in sandy areas between rocks at White's Pt., Little Corona and Laguna Beach.

229

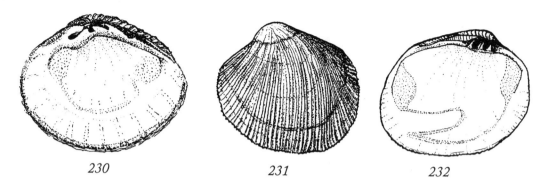

230　　　　　　*231*　　　　　　*232*

Prototbaca staminea—the littleneck clam or common littleneck clam (Figs. 231, 232). The shell has numerous, well-developed radiating ribs and few, less prominent concentric ridges. The shell is white to yellow or tan in color with or without V-shaped brown markings. The inside of the shell is white. Specimens attain lengths of 3 inches. This clam is often used in making chowders or other clam dishes. The littleneck clam typically lives in the substrate to depths of 8 inches. This species is encountered commonly in the sand and gravel under rocks in the low tide zone at the rocky shore of the Palos Verdes Peninsula, Little Corona and Laguna Beach. It also occurs in sandy outer parts of bays at Marina del Rey and Newport Bay.

Saxidomus nuttalli—the smooth Washington clam or butter clam (Figs. 233, 234). The shells are thick and heavy with numerous concentric growth ridges on the outer surface. The 2 shells gape at the posterior end where the siphon emerges. The shells are yellow-white outside and white inside except for purple markings near the posterior or siphonal end. Specimens may attain lengths of up to 7 inches. It lives within the substrate to depths of 18 inches. The smooth Washington clam is commercially harvested especially in British Columbia. It is collected locally for personal use. *Saxidomus nuttalli* has been taken from intertidal waters in the sand near the entrances of Alamitos Bay and Newport Bay.

233

Tivela stultorum—the Pismo clam (Figs. 235, 236). The shells are triangular in shape, heavy and possess a porcelain-like luster. The periostracum is generally present and has a varnish-like apearance. The shells are white and often have

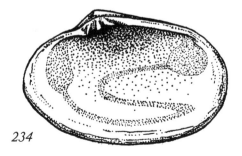

234

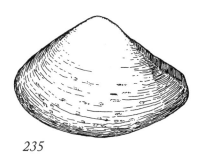

235

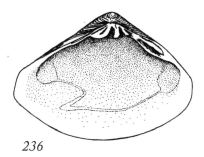

236

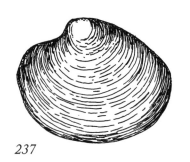

237

radiating bands of brown color. Specimens can attain widths of 6 inches. The Pismo clam makes an excellent clam chowder. Besides the human the Pismo clam is eaten by crabs, moon snails, sharks, rays, shorebirds and sea otters. The Pismo clam lives in the low tide zone along such offshore sandy beaches as Santa Monica, Long Beach, Seal Beach and Huntington Beach. This species is protected by law. Consult the local California Department of Fish & Game office when the Pismo clam can be collected legally for size limitations and the number of specimens which may be taken.

Amiantis callosa—white amiantis (Fig. 237). A large, white bivalve which measures up to 4 inches in diameter. It has many, regularly spaced concentric ribs which may branch. The hinge is well developed. The white amiantis lives in subtidal sandy beaches, but empty shells are often seen washed up on the beach. The hinge region of the shell is more commonly found.

Family Mactridae—the dish clam (Figs. 238, 239). The common name of this family is derived from the presence of a spoon-shaped pit on the inside of each shell.

Tresus nuttallii—the gaper clam (Figs. 238, 239). The shells of the gaper clam may reach lengths of 8 inches and weigh 4 pounds. A large siphon extends through a gap at the posterior end of the shell (Fig. 200, right side), hence its common name. Large specimens may live 2-3 feet beneath the surface of

238

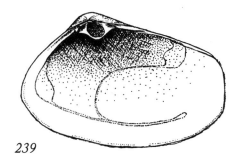

239

the substrate. The shells are brown to black in color depending upon how much of the thin, brown periostracum is present and upon the color of the substrate. The long siphon makes an excellent clam steak. Specimens have been dug, always with some difficulty, from Cabrillo Beach, Alamitos Bay and Newport Bay. This species is known also under the name *Schizothaeris nuttallii.*

Family Tellinidae—the tellin clams (Figs. 240-249). These clams have flattened shells and a ligament at the hinges which may be seen from the outside. The surface of the shells is smooth and may be colored. Teeth are present on the inside of the shell at the beak.

Florimetis obesa—**yellow clam** (Figs. 240, 241). This species can be identified by the 2 ridges at the posterior end of the shell (Fig. 240). The shell is yellowish-white both on the inside and outside of the shell. Specimens will measure up to 3 inches in width. Specimens have been taken from beneath the surface in sandy areas of Alamitos Bay and Newport Bay and also in sand under rocks at low tides at Little Corona. It was known previously as *Apolymetis biangulata.*

Macoma nasuta—**the bent nose clam** (Figs. 242, 243). This clam is so named because the shells are bent to the right. The shells are white but may be covered with a gray periostracum. Specimens seldom attain lengths greater than 2 inches. The animal lives in fine sediment to a depth of a few inches. To feed the siphon extends above the surface and sucks up the fine debris, mud and sand from the surface of the substrate. This material is transported to the digestive tract with the undigested food, mud and sand being expelled out the excurrent siphon. It is a common clam in the bays and harbors of Southern California; specifically, it has been collected at Marina del Rey, Los Angeles-Long Beach Harbors, Alamitos Bay, Anaheim Bay, Huntington Harbour and Newport Bay.

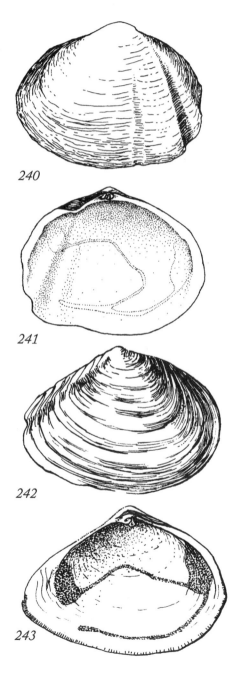

240

241

242

243

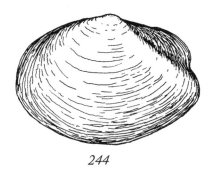

244

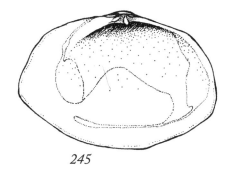

245

Macoma secta—the white sand clam (Figs. 244, 245). The shells are white both inside and outside. Some gray periostracum may persist at the posterior or siphonal end of the animal. The shells are nearly identical to one another and are not bent as in *Macoma nasuta*. Specimens may obtain lengths of 4 1/2 inches. The white sand clam feeds the same way as *Macoma nasuta*. Large specimens burrow into the sand to depths of 18 inches in the intertidal and subtidal areas of Cabrillo Beach, Alamitos Bay and Newport Bay.

Tellina buttoni (Figs. 246, 247). Specimens of this species are white and measure 1/2-3/4 inch in length. An oblique rib may be found across the inside of the anterior part of the shell. It has been collected subtidally from the outer reached of Alamitos Bay; it is known to occur subtidally in offshore waters. This species is known also under the scientific name of *Tellina modesta*.

Tellina carpeteri (Figs. 248, 249). This species is characterized by a thin pink colored shell which may have a glossy appearance. Specimens reach 1/2 inch in length. This species occurs in the same areas as *Tellina buttoni*.

Family Semelidae (Figs. 250, 251). Heavy shells possessing concentric sculpturing on the outer surface. The lateral teeth are well-developed (Fig. 251).

Semele decisa—the clipped semele (Figs. 250, 251). The common name comes from the seemingly cut margin of one side of the shell (Fig. 250, right

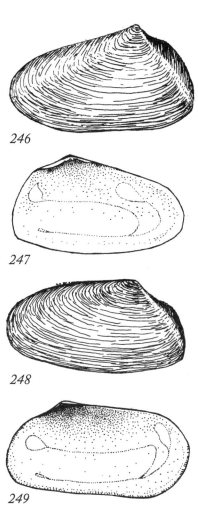

246

247

248

249

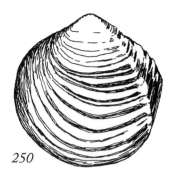

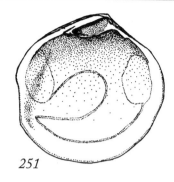

250

251

side). The shell is white with purple pigmentation around the hinge. Specimens will reach 3 1/2 inches. Living specimens have been observed in the sand under rocks at low tide at Little Corona and Laguna Beach.

Family Donacidae (Figs. 252, 253). Small, triangular clams characterized by possessing lateral teeth, cardinal teeth and an external ligament.

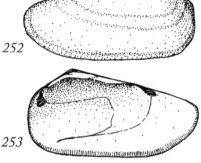

252

253

Donax gouldii—the bean clam (Figs. 252, 253). A small shell generally white to cream in color with darker markings. The lower margins of the shell are crenulated(Fig. 252). Shells measures about 3/4 inch in length. Population explosions of this species occur in the intertidal sandy beaches of Southern California. One such population explosion will generally last 2-3 years as the clams mature, then living specimens will disappear from that particular area for one or more years. The hydroid *Clytia bakeri* (Fig. 22) attaches to living specimens. The bean clam is used locally in making a clam chowder.

Family Psammobiidae—the sand clams or sunset clams (Figs. 254-259). These clams are characterized by the long external ligament.

Gari californica—sunset clam (Fig. 254). A thin, smooth shell which is yellow white in color. It measure up to 4 inches in length. A thin membrane (periostracum) is usually present at the margins. Empty shells are often seen washed up on sandy beaches. The sunset clam lives at the entrances of bays, such as Alamitos Bay, and in off shore sandy or gravely bottoms.

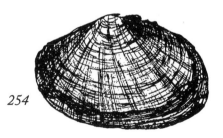

254

Nuttallia nuttalli—the purple clam (Figs. 255, 256). The shells have a purple tinge both inside and outside. The thin periostracum usually persists over the shell; it is brown and frequently appears glossy. Specimens may reach 4 inches in length.

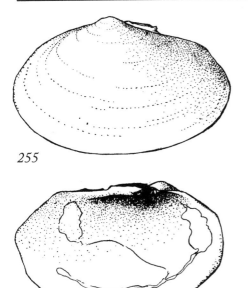

255

256

Large specimens may burrow in the sand to depths of one foot or more. Specimens have been collected intertidally from the sandy beaches of Alamitos Bay and Newport Bay. This species is also known under the scientific name *Sanguinolaria nuttalli.*

Tagelus californianus—the **California jack-knife clam** (Figs. 257, 258). The hinge is located approximately in the middle of the shell. The color is white to gray with a brown periostracum covering the shell. The Califonia jack-knife clam reaches lengths of 4 inches. It is used as fish bait in Southern California. *Tagelus californianus* is widespread both intertidally and subtidally in the softer substrates in Alamitos Bay, Anaheim Bay and Newport Bay.

Tagelus subteres (Fig. 259). This species is smaller and not as common at the California jack-knife clam. Mature specimens measure 2 inches in length. A thin, glossy membrane (periostracum) covers most of the shell. The heavier parts of the interior of the shell is stained violet which often shows through the thin, white shell. It is found in the same habitat as the California jack-knife clam.

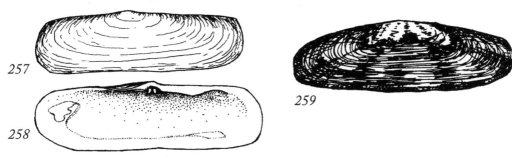

257

258

259

Family Solenidae—the **razor clams** (Figs. 260, 261). The razor clams are distinguished from the other families by having shells which gape at both ends.

Solen rosaceus—the **rosy razor clam** (Figs. 260, 261). The hinge is located at the extreme anterior end which distinguishes it from *Tagelus californianus.* The shells are rosy pink in color which shows through a thin, glossy periostracum. This species grows to lengths

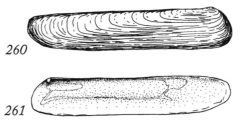

260

261

of 3 inches and can burrow in the sand to depths of one foot. A few specimens have been dredged from the bottom in Alamitos Bay and Newport Bay.

Family Hiatellidae—the boring clams (Figs. 262, 263). These clams are so named because some of the species either nestle or burrow into algal holdfasts. The shells may become quite irregular in shape because of the presence of other organisms.

Hiatella arctica (Fig. 262). The thin white shells of this species may be quite wrinkled or irregular in shape. Specimens may reach one inch in length but this is an unusual size for specimens collected locally. *Hiatella arctica* is widespread throughout the area especially within the bay mussel beds. It has been taken from Marina del Rey, King's Harbor, Los Angeles-Long Beach Harbors, Alamitos Bay, Anaheim Bay, Huntington Harbour and Newport Bay. It has also been collected from the rocky shores of Palos Verdes Peninsula from algal holdfasts at the minus tide levels. This species has also be referred to as *Saxicava arctica*.

262

263

Panopea gemersa—**geoduck** (Fig. 263). A large, white species measuring up to 7 inches in length. It burrows into the sandy mud present in bays to a depth of 2-3 feet. The large siphon extends to the surface to feed and cannot be retracted into the shell. The siphon is prized for clam chowder or a clam steak. It is not common in this environment in Southern California; it is common in the northwest.

Family Pholadidae—the pholads or piddocks (Figs. 264-269). Species of this family are capable of boring into rocks, hard packed substrates or occasionally wood. The boring is accomplished by the mechanical action of the movement of the 2 shells; the anterior margin of the shells may have serrations or teeth present (Figs. 268, 269) which assist the boring. The surface of the shells are roughened and sculptured. Rocks can be found often with circular holes which represent the action of this family.

Parapholas californica—**the scale-sided piddock** (Figs. 264, 265). This species can be readily distinguished from the other pholads by the brown periostracum which extends

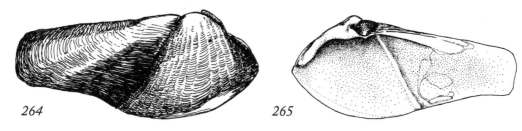

264

265

diagonally over the posterior third of the shell. Specimens reach 4 inches in length. *Parapholas californica* has been collected at White's Pt. and Pt. Fermin during low tides from soft rocks.

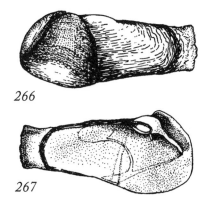

266

267

Penitella penita—the flat-tipped piddock (Figs. 266-267). The common name comes from the flattened plate which occurs just anterior to the hinge and covers over part of the shells. The periostracum extends over the posterior end of the 2 shells as a membrane. Shells measure 2-3 inches in length. *Penitella penita* has been found at all rocky shores in the Palos Verdes Peninsula at low tide where specimens have burrowed into softer rocks. It has also been taken at Dana Pt. This species has also been known under the scientific name *Pholadidae penita*.

Zirfaea pilsbryi—the rough piddock (Figs. 268, 269). The surface of this piddock has a rough texture and the anterior portion is provided with forwardly directed teeth. The 2 shells do not meet anteriorly. Specimens are known to reach 5 inches in length and burrow to depths of one foot in hard packed muds in bays. Local specimens are smaller and have been collected from rocks at White's Pt., Pt. Fermin and near the entrance of Alamitos Bay.

Family Teredinidae—the shipworms or teredos (Figs. 270, 271). The shipworms are elongated, worm-like pelecypods which live in wood. They line their burrow with a white calcareous tube which they secrete. The burrow is closed by 2 structures termed pallets (Fig. 271).

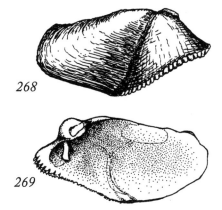

268

269

Lyrodus pedicellatus (Figs. 270, 271). This shipworm is smaller than most of the other species of teredos. Its burrow rarely extends more than 6 inches. Painted or creosoted

270

271

structures prevent larval shipworms from entering the wood; however, if a split or break occurs in the piling or if a chip of paint comes off the side of a boat, then it will be possible for teredos to enter and begin damage. *Lyrodus pedicellatus* is known to occur in Los Angeles-Long Beach Harbors and Alamitos Bay, but undoubtedly it occurs in the other protected waters of Southern California. The local damage caused by *Lyrodus pedicellatus* has been primarily within Los Angeles Harbor; it has been minimal elsewhere. This species was formerly known locally as *Teredo diegensis*. A second pelecypod wood borer, *Bankia setacea*, is known to occur in offshore pilings; it is 2-4 times larger than *Lyrodus pedicellatus*.

Family Lysoniidae (Fig. 272).   The members of this family are small and their shell thin and easily broken. The inside of the shell is pearly in appearance.

Lysonia californica—the **California lysonia** (Fig. 272).   This species rarely reaches one inch in length. The posterior end of the shell is elongated and frequently has bits of sand attached. It is only known from the subtidal regions of Alamitos Bay, but because of its small size it has probably been overlooked from other bays and harbors.

272

CLASS CEPHALOPODA—**the octopus, squid, cuttlefish, chambered nautilus** (Figs. 273-274). The cephalopods are highly developed molluscs with complex eyes much like those of vertebrates. The head bears long tentacles provided with suckers which accounts for its class name (head-footed animals). Several species of octopus are known from from Southern California and one species of squid can be encountered in local waters, especially around Catalina Island. The paper nautilus (not figured) occurs offshore from this island and its empty, fragile shells may be washed ashore.

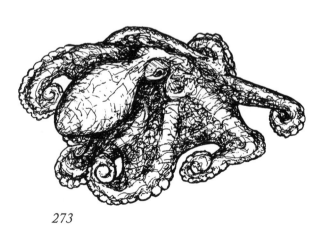

273

Octopus bimaculoides—**two-spotted octopus or the mud-flat octopus** (Fig. 273).   Two dark spots occur in front of the eyes from which the common name is derived. This species is the one most people come in contact with in shallow waters. It is found under rocks at low tide at all rocky shores of Southern California. A closely related species, *Octopus bimaculatus*, is found in deeper waters of Southern California. It is also referred to as the two-spotted octopus. Specimens of

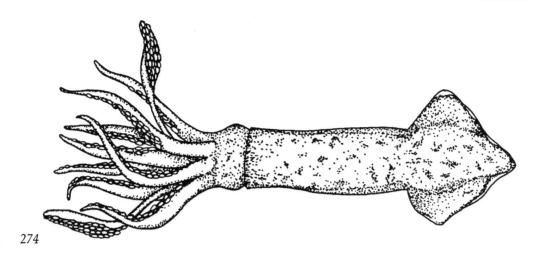

274

these species are known to reach 2 to 3 feet across but these are rare; they are usually less than a foot across. Octopus are used as food by humans, but the use of any chemical to facilitate collection is prohibited by law.

Loligo opalescens—**common squid** (Fig. 274). This pelagic cephalopod is characterized by have 8 arms and 2 tentacles. It will measure up tp 12 inches in length. Adults are found offshore and they come inshore to spawn. Spawning often occurs in submarine canyons such as the canyon off Scripps Institution of Oceanography in San Diego County. They attach egg capsules containing embryos to the sides of the canyon. Squid feed on small fish and crustaceans and in turn are fed upon by fish, birds and mammals. Squid in California were initially used as fish bait but in more recent years have been eaten by humans. Squid are attracted to night lights, and they are often observed this way off Catalina Island.

CLASS SCAPHOPODA—the tusk or tooth shells (Fig. 275). This class of mollusks is characterized a tubular shell which is open at both ends. The diameter of the shell increases with age and the larger end extends into the sediment. It feeds upon small microscopic organisms present between the sand grains.

Dentalium neohexagonum—**six-sided tooth shell** (Fig. 275). The common name is derived from the hexagonial shape in cross-section. It measures slightly over an inch in length. No tusk shells occur intertidally in Southern California and empty shells are rarely washed up onto the beach. It is present in sandy mud off shore in subtidal depths.

275

PHYLUM ARTHROPODA
The crustaceans, insects, spiders, etc.

The phylum Arthropoda is the largest animal group and comprises more than one million different species, the majority of which are insects. Members of this phylum are found in all environments on the earth. Only one group of arthropods, the crustaceans, have successfully invaded, in terms of numbers of species, the ocean environment. The arthropods are characterized by possession of a hard outside skeleton, the exoskeleton. These animals must periodically shed their skeleton, by a process known at molting, in order to grow larger. At first the new skeleton is soft then hardens. The presence of jointed appendages allows for greater adaptability in locomotion and other activities.

CLASS CRUSTACEA—The crustaceans (Figs. 276-351). The crustaceans are separated from the other arthropods by having a head region bearing 5 pairs of appendages, 2 pairs of which are antennae.

Subclass Ostracoda—The ostracods (Fig. 276). These are small animals of about 0.1 inch in length in which 2 plates fit over the animal giving it a superficial appearance to a small clam. Many species occur in the local marine environment which are difficult to identify. They are found in algal mats, algae holdfasts and in sand or mud substrates in most if not all the local collecting sites.

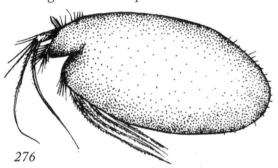

276

Subclass Copepoda—The copepods (Figs. 277-279). The copepods are small animals which play an important role in the food chain of the sea. They feed upon plant material, such as diatoms, and they in turn are fed upon by fish directly or indirectly by way of another animal. Copepods have a single median eye, a modified first antennae for swimming and 5 pairs of legs.

Order Calanoida (Fig. 277). This group of copepods may be distinguished by the possession of a very long first pair of antennae and the two body regions sharply distinct.

Calanus sp. (Fig. 277) This and related species are largely pelagic and a plankton net is necessary in order to collect them. They can be collected from offshore waters and within the entrances of bays and harbors. *Calanus* sp. is about 0.15 inches in length and white in color.

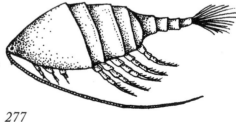

277

Order Cyclopoida (Fig. 278) The antennae are moderately long in this group and the two body regions are not sharply distict. This group of copepods is typically found in fresh water.

Pseudomyicola spinosus (Fig. 278) This species is found in the mantle cavity of the bay mussel *Mytilus edulis*. The living copepod appears white in color in contrast to the yellow-brown gills of the mussel. The mussel pumps water through its gills and filters out microscopic planktonic organisms which constitutes its food. The copepod also feeds upon these microscopic organisms. It is a commensal relationship between the two species; the copepod does not affect the mussel. This species of copepod has a high incidence of occurrence wherever the bay mussel is found. It has also been collected from the ribbed horsemussel *Geukensis demissus* (Fig. 200).

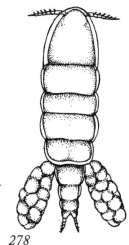

278

Order Harpacticoida (Fig. 279). The harpacticoid copepods have very short first antennae and the two body regions are not distinct.

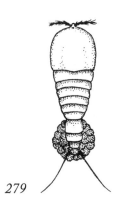

279

Tigriopus californicus—**the tide pool copepod** (Fig. 279). This copepod is a small, red-orange species which is found abundantly in small tide pools in the mid- to high tide zone in the rocky shore environment. This species can be observed in this niche by looking carefully for reddish-orange dots moving about in the tide pool. While this species of copepods can filter feed, it primarily browers the film of microscopic algae that grows in the tide pool. They are capable of withstanding great variations in water temperature and salinity.

Subclass Cirripedia—**the barnacles** (Figs. 280-289). Two general types of barnacles exist; the stalked forms (Figs. 280, 281) and the acorn forms (Figs. 282-289). Barnacles are characterized by having 6 pairs of legs which are used for filtering fine particulate matter from the water. Their body is protected by few to many calcareous plates. Barnacles attach to solid substrate during their pelagic larval life.

Order Thoracea (Figs. 280-289). All local species belong to this order which is characterized by having calcareous plates and either being free-living or commensal.

Family Scalpellidae (Fig. 280). Stalked barnacles possessing many calcareous plates.

Pollicipes polymerus—**the goose-neck barnacle** (Fig. 280). This species is characterized by having many calcareous plates of various

280

sizes. Specimens may reach 4 inches in length. The goose-neck barnacle is a common inhabitant in the intertidal zone of rocky shores where is grows in clusters especially in association with the California mussel *Mytilus californianus* (Figs. 201, 202). It has been either collected or observed at all rocky shores and the rock jetties leading into bays. The stalk of this animal has been used in the preparation of a chowder.

Lepas anatifera (Fig. 281). This stalked barnacle lives in the pelagic environment. Clusters of it attach to a variety of living and non-living floating objects. It is frequently seen washed up on sandy beaches especially after storms.

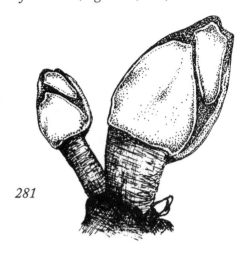

281

Family Balanidae (Figs. 282-288). Acorn barnacles in which the radial calcareous plate fits outside the lateral ones.

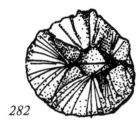

282

283

Balanus amphitrite (Figs. 282, 283). This cosmopolitan species may be distinguished from the other acorn barnacles by the presence of fine pink to purple lines along the outside of the calcareous shells. Large specimens may reach 0.25-0.5 inches in diameter. *Balanus amphitrite* is found abundantly in the high tide zone level of the pilings (Fig. 11) in all local bays and harbors. This barnacle was accidently introduced into the Salton Sea presumably from sea planes during World War II. It is attaches to roots, rocks, etc. throughout the Salton Sea. It is associated with a much smaller species; *Chthamalus fissus*, and another similar sized species, *Balanus crenatus*, which lacks the colored lines.

Balanus crenatus (Fig. 284). This species is white to gray in color with the outer shell margins rough. Local specimens are generally less than 0.5 inches in diameter. Specimens have been collected from pilings in Marina del Rey, outer Los Angeles-Long Beach Harbors, Alamitos Bay and Newport Bay. It is present with *Balanus amphitrite* in the high tide level of a piling (Fig. 11).

284

Balanus glandula (Figs. 10. 285). The small white to olive colored barnacle is one of the characteristic animals of the high tide horizon at rocky shores (Fig. 10). It occurs with the

285

smaller barnacle *Chthamalus fissus*. Specimens of *Balanus glandula* measure about 0.5 inches in diameter. They may be distinguished by the ridges which extend downward along the sides. This species is found at all Southern California rocky shores and along the rocky jetties leading into bays and harbors.

Balanus nubilis (Fig. 286). This is the largest species of barnacle on the Pacific Coast where specimens have been measured nearly 5 inches in diameter. It is not a common inhabitant of Southern California; when present, it attaches to rocks and pilings in shallow, subtidal waters. It has been reported that this species of barnacle was used for food by the Northwest native Americans.

286

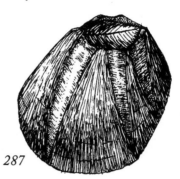

287

Megabalanus californicus—the red and white barnacle (Fig. 287). The outer shells of this barnacle are pink to red in color with fine white vertical lines. Specimens may reach up to 1 inch in diameter. The red and white barnacle is present at lower intertidal zones where it attaches to rocks. It is found at all local rocky shores and the rock jetties leading into bays. This species has also been known under the name *Balanus tintinnabulum*.

Tetraclita rubescens—the thatched barnacle (Fig. 288). This species resembles a thatched roof which accounts for its common name. The individual, outer plates in larger individuals usually cannot be seen. The surface is a darkened red color. Specimens reach 1 inch in diameter. The thatched barnacle is commonly found at rocky shores and rock jetties in the low tide zone. This species has also been known under the name *Tetraclita squamosa*.

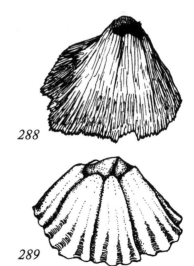

288

Family Chthamalidae (Fig. 289). Acorn barnacles in which the radial calcareous plate fits outside the lateral ones.

Chthamalus fissus—the small acorn barnacle (Figs. 10-12, 289). This small but nevertheless abundant species

289

is gray-brown in color and measures only about 0.15 inches in diameter. It is a characteristic species of the high tide zone of all local rocky shores (Fig. 10). This small barnacle feeds whenever waves splash over it or when it is covered with water during high tide. It also occurs in the outer reaches of the bays and harbors of Southern California where it attaches to pilings (Fig. 11).

Subclass Malacostraca (Figs. 290-351). This group includes most of the large well-known crustaceans such as the crabs and lobsters. They may be distinguished from the other crustaceans by their compound eyes and distinct thoracic and abdominal regions.

Order Leptostraca (Fig. 290). This group is known from only about 10 species. The leptostracans possess 7 abdominal segments rather than 6 which is characteristic for the other types of malacostracans.

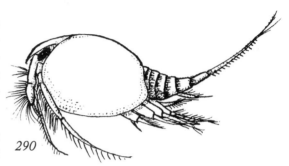

Epinebalia **sp.** (Fig. 290). This species has 2 lateral plates which fit over most of the animal's body except part of the abdominal region. Mature specimens measure about 0.75 inches in length. This species occurs abundantly in local bodies of water which are polluted such as the inner reaches of Los Angeles-Long Beach Harbors. *Epinebalia* is primarily a filter feeder, but it probably feeds also on small bits of decaying plants. It has been obtained from Alamitos Bay from jars, which contained some commercial fertilizer, suspended in the subtidal water for a month.

Order Mysidacea (Fig. 291). Mysids resemble small shrimp, but they differ is some anatomical details including the last thoracic segments are not fused with the anterior ones or the head as in the true shrimp. Mysids are both marine and fresh water inhabitants. In the marine environment they live in the water column or crawl on the surface of the sediment.

Holmesimysis costata (Fig. 291). This species is the most common of the local mysids. It lives in association with the giant kelp (Fig. 540) on which it feeds. Mysids are important food items for many species of fish.

Order Cumacea (Fig. 292). Superficially the cumaceans resemble the leptostracans but their lateral plates only extend to 3-4 thoracic segment rather than mid-way through the abdominal region.

291

104

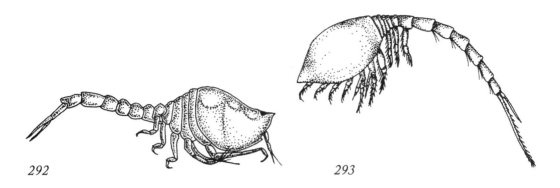

292   *293*

Cumella vulgaris (Fig. 292). This is a small (0.1 inch long) but very common cumacean. It differs from the other cumaceans by its black color. During the day it lives in the sediment around the base of the giant kelp; at night it migrates upward to the surface of the water. Collections of this species are made with a plankton net at night.

Oxyurostylis pacifica (Fig. 293). This species measures from 0.25-0.5 inches in length and is white to tan in color. This species lives on the surface of the mud both intertidally and subtidally in Alamitos Bay and Newport Bay.

Order Isopoda (Figs. 293-301). The word isopoda means similar feet which refers to the characteristic feature of having 7 pairs of similar thoracic legs. The majority of the isopods have a body which is flattened dorso-ventrally.

Paracerceis gilliana (Fig. 294). This isopod resembles a sow bug found in backyard gardens. It is dark gray in color and measures up to 0.5 inch in length. The exoskeleton is heavy and has many bumps along the upper surface. *Paracerceis gilliana* is typically found within clumps of the bay mussel attached to pilings at Marina del Rey, Alamitos Bay and Newport Bay.

294   *295*

Eurydice caudata (Fig. 295). This isopod resembles *Cirolana harfordi* (Fig. 296), but is much smaller measuring only 0.25 of an inch. It occurs offshore under similar conditions as *Cumella vulgaris* (Fig. 292). It is collected at night with a plankton net.

Cirolana harfordi (Fig. 296). This gray colored isopod measures up to 0.75 inches in length. It is found abundantly along both sandy and rocky shores throughout Southern California. It is especially abundant in mussel beds. It can be seen at low tide in the sand at sandy beaches or among washed up algae. It is present under rocks or in the rubble

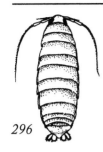

beneath the rock found in the low tide zone. This species feeds on a variety of foods; it is an important scavanger and is capable of cleaning a dead fish or other animals. It also is a predator feeding on polychaetes and amphipods.

Idotea sp. (Fig. 297). This species of isopod may reach up to 1.5 inches in length. It can be identified by its long abdominal region, its narrow body and by its olive-green coloring. It is collected from algae, typically the larger brown species, from the low tide horizons at rocky shores. It feeds on brown algae and may be difficult to see since it is the same color as the algae on which it lives. It is fed upon by many species of fish. An occasional specimen may be found among algae washed upon sandy beaches.

296

297

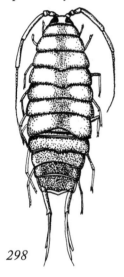

Ligia occidentalis—the rock louse (Fig. 298). This gray colored isopod will measure up to 1 inch in length. It scurries about the rocks in the high tide zone at all such areas; it is the largest and most frequently seen local isopod and a characteristic animal of the splash zone. The rock louse is especially abundant on the rock jetties where people fish; they feed upon the food, fish bait and fish left by the people.

Limnoria tripunctata—the gribble (Figs. 299, 300). This small, gray isopod measures about 0.1 inch in length. It burrows into wood structures especially into the soft or spring wood (Fig. 300). The burrows run parallel to the surface and at intervals a small hole is opened to the outside to facilitate water circulation. The female broods the young on the under surface of her thoracic region; they hatch out as small, adult-like animals. The young either reinfest the same piece of wood or migrate through the water to another piece of wood. As discussed in the Introduction, this species of gribble is capable of burrowing through creosoted wood. *Limnoria tripunctata* is found throughout Southern California wherever wooden structures are found; however, its heaviest infestation is in Los Angeles Harbor. Specimens of this species have either been observed or collected on suspended wooden blocks from Marina del Rey, Los Angeles-Long Beach Harbors, Alamitos Bay, Anaheim Bay, Huntington Harbour and Newport Bay.

298

299

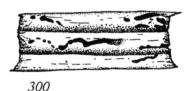

300

Cyanthura munda (Fig. 301). This long, slender gray isopod measusres up to 0.75 inches in length. It has been collected from *Mytilus edulis* (Fig. 203) attached to boat floats in Los Angeles Harbor, Alamitos Bay and Newport Bay. It probably feeds on the smaller species of algae and organic material attached to the mussels. It can bite humans especially in the softer skin between the fingers; its bite can inflct some pain and irritation to the human.

√ **Order Tanaidacea—the tanaids** (Fig. 302). This group of animals resemble isopods; they differ from them in that the first pair of thoracic legs have pincers (gnathopods).

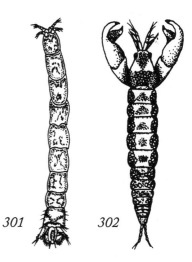

301   302

Anatanais normani (Fig. 302). This brown and whitish tanaid measures about 0.3 inches in length and is the most common intertidal species of this group. It is found among algal holdfasts at the rocky shores of the Palos Verdes Peninsula and on boat floats at King's Harbor and Newport Bay.

Order Amphipoda (Figs. 303-314). The term amphipoda refers to the possession of two types of thoracic legs. There are 2 pairs of mit-like legs, known as gnathopods (for example, Figs. 303, 307)), and 5 pairs of non-modified thoracic legs. Another characteristic feature, which separates the amphipods from the isopods and tanaids, is that the body is usually flattened laterally instead of dorso-ventrally. Four general groups of amphipods occur but only three types, the gammarideans (Figs. 303-312), the caprellids (Fig. 313) and the hyperiids (Fig. 314) are discussed herein.

Ampithoe plumulosa (Fig. 303). This flesh-colored amphipod has many small pigmented spots over the body, especially on the upper surface. Specimens will reach 0.5 inches in length. The second pair of antennae possess many fine bristles along the inner margin. *Ampithoe plumulosa* is especially abundant during the spring and summer months on pilings and boat floats in Marina del Rey, Alamitos Bay and Newport Bay. Occasional specimens have been observed in coralline algae (Figs. 554-557) and surf grass (Fig. 441) at Little Corona.

Aorides columbiae (Fig. 304). This species measures about 0.3 inches in length. It can be identified by the presence of a large tooth present on

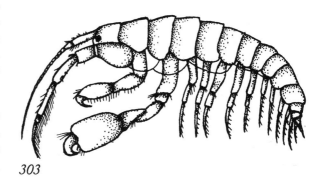

303

107

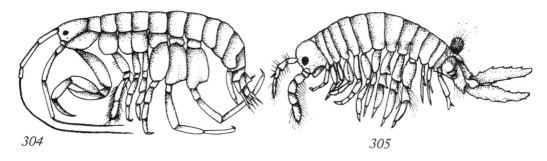

304   305

the first thoracic leg (Fig. 304) of the male; however, magnification is necessary in order to see this characteristic. *Aorides columbiae* is known to be common from surf grass and coralline algae at Little Corona; its presence has not been checked elsewhere but it is probably found at all rocky shores at low tide.

Chelura terebrans (Fig. 305). A wood boring amphipod which occurs with *Limnoria tripunctata* (Fig. 299). Large specimens measure about 0.75 inches in length and can be identified by its light color and the elongated development of the last pair of appendages (Fig. 305). This species is known locally from only Los Angeles Harbor where it causes little or no damage to wooden structures.

✔ *Corophium acherusicum* (Fig. 306). This species of amphipod is deeply colored with a brown pigment over the upper surface and on parts of the legs. Large specimens measure up to 0.3 inches. The second antennae are ex-tremely large, and the posterior end of the body is flattened. *Corophium acherusicum* con-structs simple mud tubes on pilings, boat floats, and in the mud protected waters. An-other species, *Corophium insidiosum*, may occur with *C. acherusicum*. Along with *Elasmopus bampo* (Fig. 307) these species are the most com-monly encountered amphipods at all bays and harbors of Los Angeles and Orange counties.

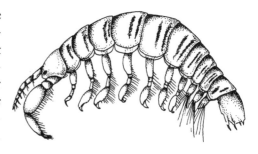

306

Elasmopus bampo (Fig. 307). This species of amphipod is tan in color and measures up to 0.3 inches in length. This species is particularly common during the summer months among the mussels attached to boat floats in protected marine waters as discussed under *Corophium acherusicum* above. This species of amphipod is easily cultured in the laboratory in aquaria with a bottom containing crushed oyster shell and fed fish flakes.

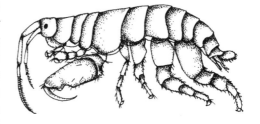

307

Hyale frequens (Fig. 308). The length of this light colored amphipod is about 0.3-0.5 inches. This species may be characterized by its large mit-like second gnathopod and its posterior thoracic lateral plates being weakly developed. Many specimens are known from species of brown and red algae and from the roots of surf grass at low tide at Pt. Dume and Little Corona. Undoubtedly this species is present from this niche at other rocky shores.

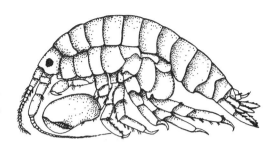

308

Jassa falcata (Fig. 309). This species of amphipod can be recognized by its reddish-tan color, its bright red eyes and the long spine at the end of the second gnathopod (Fig. 309). It measures about 0.3 inches in length. This species has been taken from a variety of habitats. It is found in clumps of bay mussels at Marina del Rey, Los Angeles-Long Beach Harbors, Alamitos Bay and Newport Bay. At Pt. Dume and Little Corona it has been taken from a variety of seaweeds and surf grass roots.

Melita sulca (Fig. 310). This species can be identified by its gray to black color and a toothed margin near the posterior end (Fig. 310). *Melita sulca* measures less than 0.5 inches in length and is found among algae and surf grass at Buff Cove, White's Pt., Pt. Fermin and Little Corona.

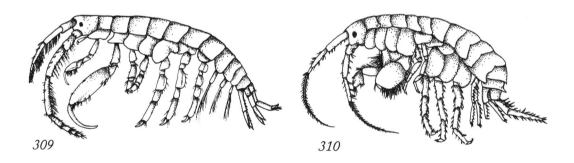

309 *310*

Orchestia traskiana (Fig. 311). This species of beach hopper is distinguished from *Orchestoidea californiana* (Fig. 312) by its smaller size and by the subchelated first pair of gnathopods in the male (Fig. 311); it is simple in the latter species. Specimens will measure about 0.5-0.75 inches in length. This flesh colored species is present on the sandy beach areas of Southern California. This species and *Orchestoidea californiana* (312) are easier to

109

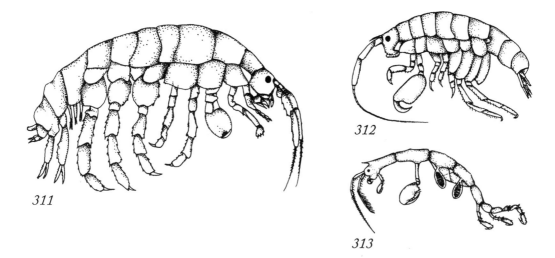

311

312

313

see at night with the aid of a flashlight; however, these two species may not be present where the beaches are raked periodically with machines.

Orchestoidea californiana (Fig. 312). This is the larger species of beach hopper which may occur with *Orchestia traskiana* (Fig. 311) as far south as Laguna Beach. It will measure over an inch in length. *Orchestia corniculata* (not figured) may also be present in the same localities. Typically a sandy beach inhabitant, it can be seen more readily at night with the aid of a flashlight as noted above.

√ *Caprella californica*—the skeleton shrimp (Fig. 313). The caprellid amphipods superficially resemble the praying mantis insect since both assume the "praying" position from time to time. Two species of the genus are commonly found in Southern California: *Caprella californica* is characterized by having a large spine on the head over the eyes (Fig. 313) and *Caprella equilibra* (not figured) lacks such a spine. Both species are tan in color with brown and red pigmented spots scattered over the body. They measure from about 0.3 to 0.75 inches in length. Both species occur on boat floats in all bays and harbors especially wherever *Obelia* (Fig. 24) or *Tubularia* (Fig. 27) are found. Under laboratory conditions the caprellid amphipods feed upon fish flakes or brine shrimp larvae.

Hyperiid amphipods (Fig. 314). This group of amphipods are marine inhabitants which either live in water column or are commensal with such animals as jellyfish (Fig. 32), ctenophores (Fig. 40) or pelagic tunicates (Fig. 416). They

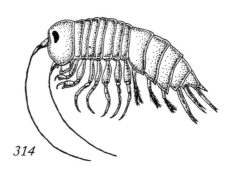

314

110

are seen more frequently in Southern California during the spring months in association with the purple striped jellyfish (Fig. 32). Hyperiid amphipods differ from the other two groups in the possession of large eyes and head (Fig. 314).

Order Decapoda—the shrimp, crayfish, lobsters and crabs (Figs. 315-361). These animals are characterized by the possession of 10 legs which arise from the thoracic region; the word decapoda means 10 legs.

Suborder Natantia—shrimps and prawns (Figs. 314-320). The body is flattened laterally; the abdominal regions and its appendages are adapted for swimming.

Family Hippolysmatidae (Figs. 315-317). This family of shrimp is characterized by the development of a long, serrated rostral spine which projects forward between the eyes.

Hippolysmata californica—the red-striped shrimp (Fig. 315). This species can be recognized by its numerous longitudinal red stripes against a light tan to cream background color. The red-striped shrimp is one of the "cleaning shrimps." It will pick up parasites and other
315
material from different species of fish; however, this is not the only source of food for this species. The red-stripped shrimp reaches about 2 inches in length and is found in tide pools in the mid- to low tide zone at many rocky shore localities of the Palos Verdes Peninsula, Little Corona and Laguna Beach. This species is also known under the name of *Lysmata californica*.

Spirontocaris picta—the green shrimp (Fig. 316). This species of shrimp, which measures an inch, has been called the green shrimp because of its color and broken back shrimp because of the right angle bend in the abdominal region. Reddish pigmented lines may be present on the thoracic region and as bars on the legs. It is commonly found in tide pools and among algae in the low tidal zone at rocky shores. It has been collected from all rocky shores visited in Los Angeles and Orange Counties.

Spirontocaris taylori—Taylor's shrimp (Fig. 317). This species of shrimp is distinguished from the others by a short rostral spine which does not extend beyond the eyes (Fig. 317).

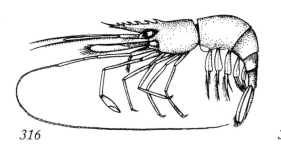

316

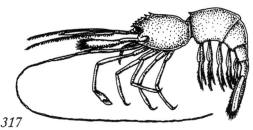

317

It is generally light in color with brown or greenish markings. It is found at rocky shores in tide pools often in association with *Hippolysmata californica* (Fig. 315).

Family Alpheidae (Fig. 318). The anterior margin of the cephalothorax (the dorsal covering over the head and thoracic region) lacks an anterior projecting spines as in some shrimp (Fig. 316) and the stalked eyes are covered with the cephalothorax.

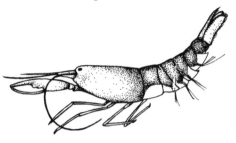

Betaeus longidactylus (Fig. 318). This species of shrimp can be recognized by the long, lower finger on the pincer which accounts for its specific name. Specimens are green to olive in color and measure about 1.5 inches in length. It inhabits tide pools in rocky shores and can be found with *Hippolysmata californica* (Fig. 315). It has been collected from the rocky shores of the Palos Verdes Peninsula, Little Corona and Laguna Beach.

318

Family Crangonidae (Figs. 319-320). Shrimp with eyes covered by the carapace and without a forward spine between the eyes. The palm of one or both first thoracic legs is large.

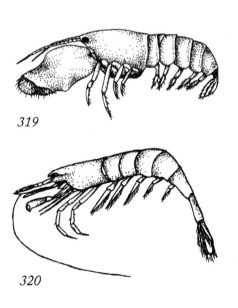

319

320

Crangon californiensis—**the pistol shrimp** (Fig. 319). This animal is so named because it is able to make a loud clicking noise with the palm and finger of its heavy first thoracic leg. This noise can be so loud that it gives the illusion of glass breaking. Indeed, it is capable of breaking the glass of an aquarium. Specimens measure 2 inches in length. The pistol shrimp is typically a subtidal inhabitant of muddy bottoms, but an occasional specimen can be collected from bays and harbors at minus tides. It has been taken at Los Angeles-Long Beach Harbors, Alamitos Bay, Anaheim Bay and Newport Bay.

Crangon nigrocauda (Fig. 320). The specific name refers to it black tail. It is found intertidally in bays such as Newport Bay among the blades of eel grass or under rocks. It is also found offshore in deeper waters.

Suborder Reptantia—the crabs and ghost shrimps (Figs. 321-332-287). The body is usually flattened dorso-ventrally. The first pair of legs usually are well developed pincers. This group, unlike the Natantia are not especially adapted for swimming.

Section Macrura. Abdominal region extended posteriorly as a crayfish.

Family Palinuridae (Fig. 321). This family can be distinguished from the others by the lack of a rostral spine.

Panulirus interruptus—**the California spiny lobster** (Fig. 321). This species of lobster is easily identified by its numerous, forward-directing spines on its thoracic region and antennae. Record specimens of the California spiny lobster will reach 3 feet and weigh 25-30 pounds. It is easily distinguished from the Maine or Eastern lobster by the absence of a large claw. The California spiny lobster is red to dark red in color. While typically a subtidal inhabitant of rocky shores, an occasional small specimen can be seen from any of the rocky shore tide pools at minus tides. They are omnivorous feeders, that is, they will

321

eat a wide variety of food either plants or animals, living or dead. Collection of the California spiny lobster is governed by law as to size, season and method of capture. Consult the nearest office of the California Department of Fish and Game for current regulations.

Family Callianassidae—the ghost shrimp (Figs. 322-324). These animals are named ghost shrimp because of their white color and because they live in burrows. The first pair of thoracic legs are unequal with one pincer much larger than the other.

Callianassa affinis (Fig. 322). This species of ghost shrimp measures about 3 inches in length and is white in color. It can be distinguished from *Callianassa californiensis* (Fig. 323) by the relative proportions of the different segments of the large first leg of the male. This species typically inhabits the sand under rocks along these shores. It has been collected from under rocks at Lunada Bay, White's Pt. and Laguna Beach.

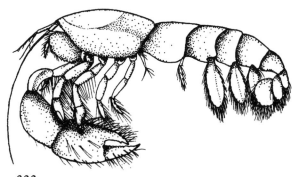

322

Callianassa californiensis—**the red ghost shrimp** (Fig. 323). This species of ghost shrimp is white with a faint tinge of red. Specimens measure up to 4 inches in length. It is very common in quiet waters of bays and harbors where it occurs both intertidally and subtidally. Ghost shrimp make extensive burrows in the sediment which serves as a home for a variety of animals including polychaetes, peanut crabs (Fig. 349) and arrow goby (Fig. 468). It has been collected, often very abundantly, from Marina del Rey, Los Angeles-Long Beach Harbors, Anaheim Bay, Huntington Harbour and Newport Bay. It is used as fishing bait by local sport fishermen.

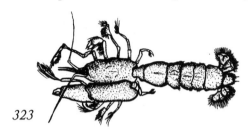

323

Upogebia pugettensis—**the blue mud shrimp** (Fig. 324). This species of ghost shrimp is not as common in Southern California as the two species of *Callianassa*. It has been collected from Anaheim Bay, but it probably also occurs in upper Newport Bay. It is distinguished from *Callianassa* by its bluish color. Its burrows are also home to many different species of invertebrates and goby fish.

Section Anomura. Abdominal region either flexed cephalothoracic region or is fleshy and twisted. The fifth pair of thoracic legs is poorly developed and often difficult to see.

Family Pauridae—**the hermit crabs** (Figs. 325-327). These comics of intertidal tide pool life are so named because of their mode of living in empty snail shells. They can be so abundant at rocky shores that it is often difficult to find an empty snail shell. The abdominal region is soft and fleshy with the last pair of appendages modified to hold the animal in the shell. The 2 pincers are unequal in size.

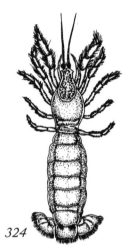

324

Isocheles pilosus (Fig. 325). The largest of our local hermit crabs; specimens may reach 3 inches in length. It is generally found occupying empty southern moon shells (Fig. 150) or channel basket shells (Fig. 174). It has been found with these shells in Alamitos Bay, Long Beach and Newport Bay during low tides. They burrow in the sediment with all but their eyes and antennae exposed. They feed on particulate organic matter. This species is also known as *Holopagurus pilosus*.

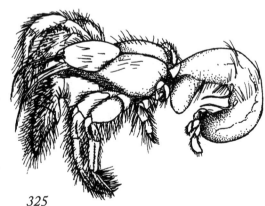

325

114

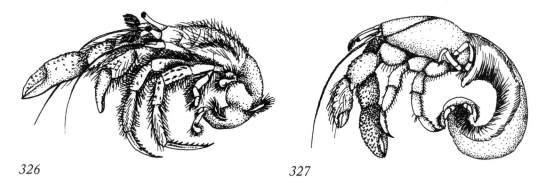

326   327

Pagurus hirsutiusculus—the hairy hermit crab (Fig. 326).   The upper surface of the thoracic region contains many stiff bristles which accounts for its common name. Specimens in Southern California tide pools are typically 1-1.5 inches in length. Bands of white or red may be seen on the appendages of smaller specimens. The hairy hermit crab inhabits a variety of snail shells especially turban shells (Figs. 121-128) and angular unicorn shells (Fig. 167). It has been collected from the jetty at Alamitos Bay, Little Corona and Laguna Beach.

Pagurus samuelis—the blue-clawed hermit crab (Fig. 327).   This small hermit crab seldom measures over an inch. It is easily identified by the bright blue color at the ends of the legs. This is the most common species of hermit crab along all rocky shores of Southern California especially in the mid-tide horizon tide pools. It is the species of hermit crab that most people see. Young specimens will inhabit a variety of small snail shells and even pieces of the scaley worm shell (Fig. 136). Adult specimens are generally found in turban shells (Figs. 121-128).

Family Hippidae—sand crabs (Fig. 328).   A common intertidal inhabitant of all sandy beaches in Southern California, especially along the Santa Monica Beach.

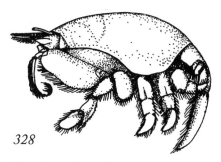

Emerita analoga—the **sand crab or mole crab** (Fig. 328).   This species is gray colored and large specimens will exceed 2 inches. The surf moves the animal up and down the beaches with each wave. It legs and uropods are modified for rapid digging in the sand. It burrows in backwards leaving its antennae exposed for feeding and respiration. Specimens are used as fish bait especially for surf fishes. The sand crab is fed upon by many species of shorebirds and surf fish.

328

Larval stages of parasites are in the body cavity of the sand crab; the adult stage of the parasite is present in shorebirds. It is found at all coastal sandy beaches.

Family Albuneidae (Fig. 329). Carapace provided with forward directing spines and the first pair of legs with pincers.

Blepharipoda occidentalis—**the spiny sand crab** (Fig. 329). A much larger and rarer species than *Emerita analoga* with which it may occur. It may reach 3-4 inches in length. It is gray in color and has habits much like the common sand crab. It is also fed upon by birds and fish. Specimens have been collected at sandy beaches during low tide at Santa Monica, Huntington Beach and at the small sandy beach at Little Corona.

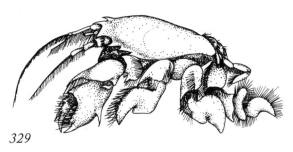

329

Family Porcellanidae—**the porcelain crabs** (Figs. 330-331). These crabs are so named because of the porcelain-like appearance of the carapace of some species. This family can be readily distinguished from the true crabs by having only one pair of antennae between the eyes instead of 2 (compare Figs. 330, 331 with Fig. 346) and the flexed fifth pair of legs (Fig. 331).

Pachycheles rudis—**the thick clawed porcellain crab** (Fig. 330). The large claws which are hairy and rough are the characters which aid in its identification. The carapace is about 0.5 to 0.75 inches across. It is light brown in color. This species of crab feeds upon small bits of organic matter and plankton. It is common in protected niches at rocky shores especially among beds of the California mussel or under rocks in the mid-tide level. It has been observed at Little Corona and Laguna Beach.

Petrolisthes cinctipes—**the porcelain crab** (Fig. 331). Sometimes called the flat procelain crab because of its flattened body. The body is smooth with hairs only at the ends of the 3 similar pairs of legs. The body of this brown colored porcelain crab measures about 0.5 inches across. The porcelain crab feeds upon small pieces of organic matter which it filters

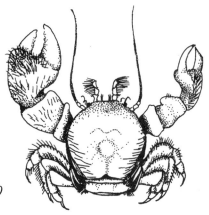

330

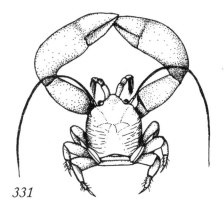

331

out from the water. It is commonly found along such rocky shores as Alamitos Bay jetty, Little Corona and Laguna Beach in beds of the California mussel or under rocks at low tides. It may be found with *Pachycheles rudis* (Fig. 330).

Family Galatheidae (Fig. 332). Body is shrimp-like; the abdomen is bent upon itself but not folded up. The tail fan is well developed for swimming. The first pair of legs is chelated, long and slender.

Pleuroncodes planipes—the **pelagic red crab** (Fig. 332). As the common name indicates this is a pelagic species which is red in color; however, it does crawl on the bottom. Southern California is generally considered to be the northern limit of its distribution, but beach strandings have been reported as far north as Monterey. The occurrence of the pelagic red crab is rare north of Baja California except when the water is warmer during the effects of El Niño. During this time it is present offshore in the countless numbers and many are washed up on the beach. The pelagic red crab provides an important component in the diet of many fish and marine mammals during the El Niño condition.

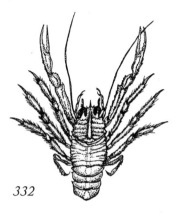

332

Section Brachyura—the true crabs (Figs. 333-351). Both pairs of antennae located between the eyes and all 5 pairs of legs well developed.

Family Leucosiidae (Fig. 333). The body is rounded, the antennae are short, and the posterior margin of the body has short spines.

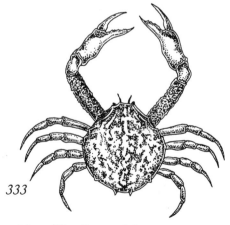

333

Randallia ornata—the **purple crab** (Fig. 333). The common name is derived from the purple or reddish mark on the upper surface. The body is more-or-less circular in shape with 2 small spines at the posterior margin. It is not a common species in Southern California which is the northern limits of its distribution. It is found in low intertidal waters buried in the sand, for example, at Little Corona.

Family Inachidae—the spider crabs (Fig. 334-337). The members of this family are so called because their long, slender legs resemble those of spiders. The front of the carapace extends forwards from the eyes.

Pugettia producta—the **kelp crab** (Fig. 334). This crab is easily identified by its olive green to brown color and the smooth upper surface. The body measures about 2-3 inches across

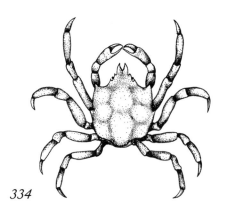

334

335

and with the legs included about 6 inches. It is present in the rocky intertidal environment at low tides where it is associated with the larger brown algae at rocky shores; largest specimens of the kelp crab occur subtidally.

Loxorhynchus crispatus—the **masking or moss crab** (Fig. 335). This species of crab is a slow moving one and the upper surface is covered with small "hairs" along with attached hydroids (Figs. 24, 26), sponges, different species of algae and other organisms. The

common name stems from the presence of these attached organisms. They eat of variety of invertebrates and seaweed. The crab measures 3 to 4 inches across. It is usually not seen in the intertidal environment, but the masking crab typically lives subtidally on piling both inshore and offshore.

Loxorhynchus grandis—**sheep crab** (Fig. 336). This species is larger than *L. crispatus* with some specimens measuring 5 inches across. Young specimens decorate themselves in much the same way as *L. crispatus*, but older specimens lack them. The sheep crab is a carnivore and scavanger. This species rarely occurs in the intertidal zone

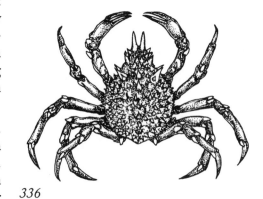

336

Heterocrypta occidentalis—**the elbow crab** (Fig. 337). The common name is derived from its long first pair of legs (containing the pincers) with a decided bent joint. The body shape is triangular which is distinctive in the true crabs. It is a subtidal inhabitant and lives among the blades of eel grass.

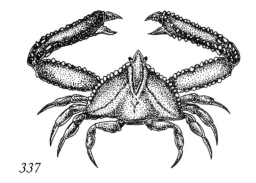

337

118

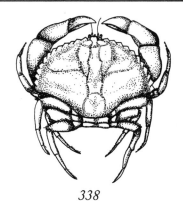

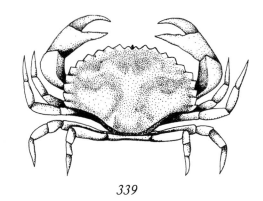

338 *339*

Family Cancridae (Figs. 338-341). The front margin of these crabs contains many serrated teeth. Members of this family are highly esteemed as food.

Cancer antennarius—**the common rock crab** (Fig. 338). The upper surface is orange-brown to dark red in color; the undersurface has many red spots on a light background. The claws are black at the tips. It is both a scavenger and a predator; it has been observed feeding upon hermit crabs. This species reaches 6 inches in width. The rock crab is found in intertidal waters more frequently along the rocks of jetties at the entrances of bays and harbors and generally partially buried in the sand. An occasional specimen has been found on the floating log-boat docks in outer Los Angeles Harbor. This species is used for food to a limited extent.

Cancer anthonyi—**the rock or yellow crab** (Fig. 339). This species is one of the smaller cancer crabs; it measures about 2.5-3 inches wide. The body is dark red and lacks spots on the undersurface. It has been found at Pt. Fermin under rocks in the low tide horizon. It also inhabits crevices in rock jetties throughout Southern California. It also eaten by humans.

Cancer magister—**the dungeness crab** (Fig. 340). This is principle commercial edible crab of the Pacific coast especially in the northwest. Large specimens measure up to 10 inches

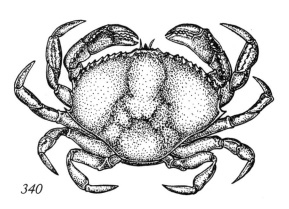

340

across. The dungeness crab is carnivorous; it is known to feed upon crustaceans of all kinds, clams, worms and small fish. It occurs in bays and offshore waters, but it is rare in Southern Californa. Some specimens are taken at low tide in bays north of Point Conception, but, when present, it is subtidal in Southern California. All species of the genus *Cancer* are governed by the Fish and Game Code of the California Department of Fish and Game. Consult the nearest office of this agency for permits.

Cancer productus—**the red crab** (Fig. 341). The red crab is often used as food since specimens will measure 6 inches across. As its common name suggests, it is bright red in color. The teeth between the eyes are 5 in number, equal in size and project forward. The red crab is a carnivore feeding upon a variety of invertebrates including barnacles. *Cancer productus* has been collected from near rock jetties burried in muddy sand at Cabrillo Beach in Los Angeles Harbor and Alamitos Bay. It has also been seen along the boat floats and rocks within marinas in Los Angeles Harbor.

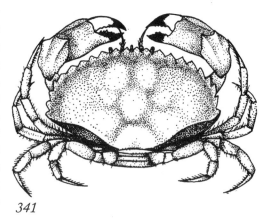

341

Family Portunidae (Fig. 342). The fifth pair of legs is modified for swimming which is flattened dorso-ventrally and is very thin in cross-section.

Portunus xantusi—**the swimming crab** (Fig. 342). The fifth pair of legs is flattened for swimming. It may be seen swimming or crawling on the bottom among eel grass. It is attracted to lights from boats at night. It will measure 7 inches across. The swimming crab is carnivorous and is known to feed upon the mole crab (Fig. 328).

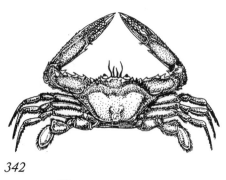

342

Family Grapsidae (Figs. 343-345). The front and back margins of the shell are more or less parallel; the side margins are parallel and generally have 1-2 teeth at the front corners.

Hemigrapsus oregonensis—**the yellow shore-crab or mud crab** (Fig. 343). This species of crab is yellow to tan in color with darker, mottled spots on the shell and legs. The legs are hairy in appearance. Large specimens may measure up to 2 inches in width. During low tides it is generally found under rocks. It is a scavenger in its feeding habits, but it will also feed on green algae and plant debris. The yellow shore-crab is found in the protected waters of back bays. It will dig into the sides of mud banks and burrow down to the water level or it lives under rocks in the intertidal zone. *Hemigrapsus oregonensis*

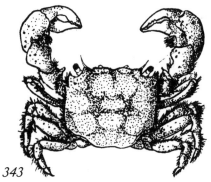

343

344

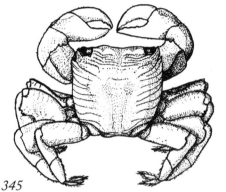

345

is known to occur in Alamitos Bay, Anaheim Bay and Newport Bay.

Hemigrapsus nudus—the **purple shore crab** (Fig. 344). This species is easily distinguished from the other grapsid crabs by the presence of purple spots on the first pair of legs (that pair bearing the pinchers). It measures 3 to 4 inches across. Its food habits are similar to *Hemigrapsus oregonensis*. It not commonly found in Southern California and where it does, it is generally found with *Pachygrapsus crassipes* (Fig. 345). It occurs at intertidal rocky shores.

Pachygrapsus crassipes—the **striped shore-crab** (Fig. 345). The upper surface is dark green in color with lighter lines running from side to side. The legs are dark purple in color. Specimens will reach 3 inches in width. This is the most commonly encountered crab along the rocky intertidal shores of Southern California. These animals scurry about the rocks and tide pools in the high rocks and tide pools in the high tide zone. They feed upon seaweeds, decaying organic matter or plant and animal material they scrape from rocks. They are found at all rocky shores and jetties in the Los Angeles-Orange Counties area; they extend into all the local bays and harbors and are found on boat docks or along the sea walls of marinas.

Family Xanthidae (Figs. 346-348). Species of this family can be recognized by the few sharp teeth on the lateral margins.

Lophopanopeus diegensis (Fig. 346). A small crab less than 1 inch in width in which the width is greater than the length. The hand of the pincers is covered with bumps. It has been found under rocks at low tides along the rocky shores of the Palos Verdes Peninsula.

Lophopanopeus leucomanus (Fig. 347). Width of the body is less than 1 inch. The length exceeds the width. The surface of the first pair of legs have irregular depressions and ridges. The tips of the pincers are black. This crab has been seen at Lunada Bay, Little Corona and Laguna Beach at low tides from under rocks.

Pilumnus spinohirsutus—**hairy crab** (Fig. 348). This small crab, which measures 1.5 inches across, is easily distinguished by its numerous hairs and spines. It usually is found buried

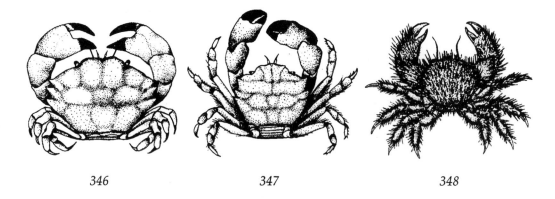

346   347   348

in sand under rocks. Its light-brown color makes it difficult to see. The hairy crab, sometimes called the retiring southerner, is found at rocky shores at low tide.

Family Pinnotheridae—the pea crabs (Figs. 349-350). Members of this family are commensal in the gill chambers of pelecypods, shrimp burrows, worm tubes, etc. They feed upon bits of food caught in the gill chamber of the clam or in the burrows of animals, but the crab does not damage the host. These crabs are small in size and have poorly developed eyes.

Pinnixa franciscana (Fig. 349). This species of pea crab is found in the burrows made by the red ghost shrimp (Fig. 323). It measures about 0.5 inch in length and is tan in color. *Pinnixa franciscana* is found wherever the red ghost shrimp is taken both intertidally and subtidally. Specific localities include Marina del Rey, Los Angeles-Long Beach Harbors, Alamitos Bay, Anaheim Bay and Newport Bay.

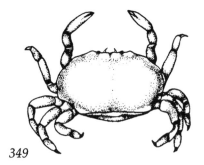

349

Fabia subquadrata (Fig. 350). This commensal crab is commensal in many species of clams including 2 species of mussels (Figs. 201-204) the most common hosts. Usually only one crab is found in a clam. The crab feeds upon the mucus trapped food collected by the clam by its filter feeding food habits. This small crab measures about 0.5 inches across. This commensal crab is found wherever the mussel is present; however, only a very small percentage of the mussels contain the crab.

350

Family Ocypodidae—the fiddler crabs (Fig. 351). The shell is more or less square in shape and without teeth. One of the first pair of legs of the male is large (Fig.287); they are of equal size in the female.

122

351

Uca crenulata—**the fiddler crab** (Fig. 351). These crabs are so named because of the behavior of the male with its large cheliped during breeding season. The male lifts this leg high into the air and then back to the body which suggests the movements of a violinist. These tan colored crabs may measure 1 inch in width. They burrow into the sand and mud above the water line in quiet waters of bays. The population of the fiddler crab has been reduced in Alamitos Bay and Newport Bay in recent years because of dredging and marina construction.

CLASS INSECTA (Figs. 352-356). Insects are the most abundant group of animals in terms of species, but only a small number of them live in the marine environment and these essentially occur in the intertidal environment. Marine insects are usually found associated with algae which had been washed up on shore. The majority of the species belong to the orders Hemiptera (true bugs), Coleoptera (beetles) and Diptera (flies). Representatives are found in some other orders including Collembola (springtails).

Anurida maritina—**the seashore springtail** (Fig. 352) This insect is a member of the Collembola group. It measures about 0.5 inches in length. The seashore springtail occurs along the Pacific Coast where often large numbers are found on the surface of small intertidal pools.

Thallassotreches barbarae—**ground beetle** (Fig. 353). This black beetle belongs to the order Coleoptera. It measures less than 0.5 inches in length and is found in rock crevices in the high intertidal rocky areas. It feeds at dusk and night on plant and animal debris. This species was first found and described from Santa Barbara which accounts for its specific name.

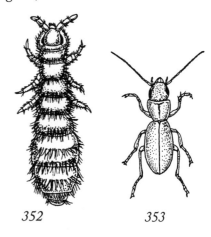

352   *353*

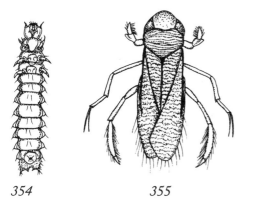

354   *355*

Aegialites californiae larvae (Fig. 354). This insect belongs to the beetle order Coleoptera. It is black and long-legged measuring up to 1 inch in length. This beetle spends its entire life in rock crevices and cracks just belong the high tide mark. The larvae are the most frequently observed stage.

Trichocorixa retriculata (Fig. 355). This insect is a true bug belonging to the order Hemiptera. It measures about 0.5 inches in length and is found in backish and saline water along the Pacific Coast.

123

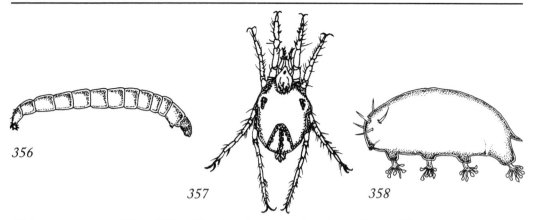

356

357   *358*

Chironomid larvae (Fig. 356). The larval stages of midges, Order Diptera, are common just beneath the surface of sand and mud in quiet marine waters. The larvae are found in the high tide horizon in all bays and harbors of Southern California.

CLASS ARACHNOIDEA—spiders, ticks, mites (Figs 357, 358). The body is divided into two region with the anterior one bearing 4 pairs of legs.

Marine mites (Fig. 357). As the name suggests, mites are extremely small measuring less than a fraction of a millimeter in length. They are common in the marine environment, but because of their size they are overlooked. They can be seen in living condition by examining the base or holdfasts of branched algae such a coralline species under a dissecting microscope. Marine mites of Southern California are poorly known.

Marine tardigrades (Fig. 358). Many species of tardigrades occur in the marine environment. Because they are so small, measuring less than 0.1 of an inch, they are virtually unknown in Southern California. They have 4 pairs of legs with the anterior 3 pairs directed forward and the fourth pair directed posteriorly. They are free living among sediment particles at sandy beaches in the high tide horizon. In some cases they may live 4 feet below the surface. Some tardigrades are parasitic in echinoderms, mussels and insects. They are also found in fresh water and damp soil.

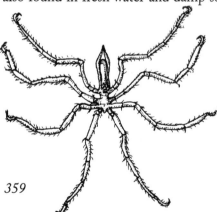

359

CLASS PYCNOGONIDAE—the sea spiders (Fig. 359). The sea spiders are characterized by a small body and from 8-12 long, slender legs. Specimens measure about 0.5-0.65 inch in width including the legs. Several species are known to occur in Southern California, but only a specialist can identify the various species. Sea spiders can be found in protected niches along the rocky shores and jetties of Southern California especially in algal holdfasts, mussel beds, clumps of ectoprocts or under rocks.

124

PHYLUM ECTOPROCTA

The word "ectoprocta" literally means outside anus, that is, to say that the anus is located outside a circle of tentacles. These tentacles surround the mouth and are used for feeding. The Ectoprocta, along with another group of animals, were formerly referred to as the Bryozoa or moss animals. This common name comes from the superficial resemblance of some of the members to mosses. The ectoprocts are colonial animals which are composed of from a few to countless numbers of microscopic individuals (Figs. 361, 363, 367, 371, 372). Some of these animals may be mistaken for red seaweed such as *Bugula neritina* (Fig. 362). The growth form of these animals may be either erect (Figs. 360, 362, 364) or encrusting over rocks, pilings, or other animals (Figs. 369, 370, 374). Some calcium carbonate is deposited in many of the erect forms (Figs. 364-366) and all encrusting forms. *Zoobotryton verticillatum* (Fig. 368) is gelatinous-like in appearance.

Bugula californica (Figs. 360, 361). This small erect species is white in color and may reach 1 inch in length. This species is of particular interest to view a living colony under the compound microscope to see the movements of the avicularia (the bird-like structures in Fig. 361). The lower jaw of the avicularia opens and closes which functions in keeping larval stages of other animals from settling upon the colony. This species in not as common as *Bugula neritina* (see below); it attaches to the lower margins of boat floats and to suspended ropes in Marina del Rey, Los Angeles-Long Beach Harbors, Alamitos Bay, Huntington Harbour, and Newport Bay.

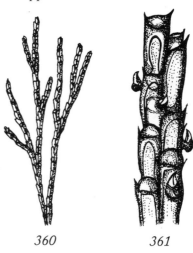

360   *361*

362   *363*

Bugula neritina (Figs. 362, 363). This species is readily distinguished by its red to purple color in life and by the absence of avicularia; the colonies of *Bugula neritina* become extensively branched and they may reach 2 inches in length. Because of its red color it may be confused with some species of red algae which are associated with it; however, examination with a hand lens will assist in separating *Bugula neritina* from the red algae (compare Figs. 362 with Fig. 569). This species of ectoproct is the most commonly encountered

125

species in protected waters; it has been collected from floating docks, boats and pilings from all bays and harbors and marinas in the Los Angeles-Orange County area.

Crisulipora occidentalis (Fig. 364).   This species of ectoproct is white, stiff and calcareous in nature. The individual zooids project outward from the main stem. The colony sometimes will reach over one inch in length. This species attaches to rocks in the low tide at most rocky shores in Southern California. It is abundant on the jetty at the entrance of Alamitos Bay.

364

Diaporoecia californica (Fig. 365).   A white calcareous species similar in appearance to *Crisulipora occidentalis* (Fig. 364). The individual zooids often cluster together in groups of 4-5. The colony may reach 1-2 inches in height. It is present on intertidal and subtidal rocks at most rocky shores in Southern California.

Phidolopora pacifica (Fig. 366).   This erect, white calcareous species can measure up to 4-5 inches in length. It is frequently branched with the branches curled slightly. Perforations in the colony generally occur as shown in Fig. 366. Colonies are found at low tides attached to the sides of rocks such as occur on the rock jetties at Alamitos Bay and Newport Bay or along the walls in small caves.

Scrupocellaria bertholeti (Fig. 367).   An erect species of ectoproct which is brittle because of the deposition of calcium carbonate in the wall of the colony. It is white in color and measures about one inch in length; it can be readily distinguished from *Bugula californica* because of its calcareous nature. Each individual has long spines which are shown in Fig. 367. This species is commonly encountered on boat floats and pilings in all bays, harbors and marinas in the local area.

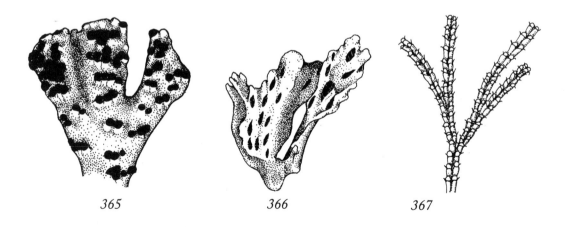

365   *366*   *367*

Thalamoporella californica (Fig. 368). This white colored, calcareous colony may form large branching structures which reach 5 inches in diameter. The branches are divided into nodes (joints) and internodes (between joints). Chitinous material, instead of calcareous deposits, is present at the nodes which gives some flexibility to the colony. It has been taken from the rock jetty at Alamitos Bay and from rocks at Little Corona.

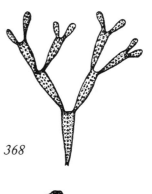

368

Holoporella brunnea (Fig. 369). This encrusting species of ectoproct is dark gray in color. Individual colonies may reach 1-2 inches in diameter; frequently adjacent colonies may more or less grow into one another. This species can be identified by the random arrangement of the individuals of the colony; however, a hand lens or dissecting microscope is necessary. This species is the most commonly encountered encrusting type found in bays. It attaches to mussels or to floats in outer Los Angeles Harbor, Alamitos Bay, Huntington Harbour and Newport Bay.

369

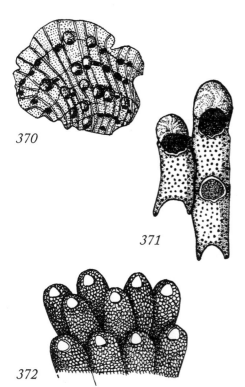

370

371

372

Schizoporella unicornis (Figs. 370, 371). The colony will grow to 1 inch or so in diameter; the color is white to gray with an orange tinge. This species can easily be mistaken for *Cryptosula pallasianna* (Fig. 372); a dissecting microscope is needed in order to make positive identification. This species is typically found encrusting over pelecypod shells such as *Mytilus edulis* in protected waters of bays, harbors and marinas. Elsewhere in Southern California it has been collected at Alamitos Bay, Huntington Harbour, and Newport Bay.

Cryptosula pallasiana (Fig. 372). This encrusting species is gray in color with an orange tinge. A colony may on occasion measure up to 1 inch in diameter. The margins of the colony may grow upward off the surface giving it a frilled appearance. It has been collected from rocks exposed at low tide from King's Harbor, Alamitos Bay, and Newport Bay.

Zoobotryton verticillatum (Fig. 373). This massive colonial species has an erect growth form and is gelatinous-like in appearance. This species has only been observed during the summer months of certain years when the water temperature was approximately 72°F, or over for a period of time. Under these conditions the colonies grew as much as 5 feet in length and 1-2 feet in width in a month or less. They have been found in Alamitos Bay, Huntington Harbour and Newport Bay attached to the undersurface of boat floats and growing downward.

Membranipora membranacea (Fig. 374). This flattened, en-crusting, white species of ectoproct is frequently found grow-

373

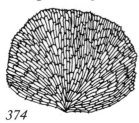

ing on the surface of the larger brown seaweeds such as *Macrocystic pyrifera* (Fig. 540) and *Egregia laevigata* (Fig. 542). The individuals of the colony are rectangular in shape which separates it from the closely related species *Membranipora villosa* has been found along the rocky shores wherever these brown algae are seen; it has been taken also near the entrances of King's Harbor and Alamitos Bay.

374

PHYLUM BRACHIOPODA
The lamp shells

The brachiopods, or lamp shells, are so named because the lower shell of some species (not shown) resemble Aladdin's lamp. The shells are attached to each other dorsoventrally rather than right and left as in the pelecypods. The internal anatomy of the brachiopods and pelecypods is vastly different. There are two general body forms in the brachiopods, the inarticulate (Fig. 375) from which a long stem, or peduncle, extends into the substrate and the articulate species which has a small peduncle extending out through an opening in the shell and attaching to rocks. All species of brachiopods in Southern California are subtidal.

Glottida albida (Fig. 375). A small inarticulate species which measures about 3 inches in length. The shells are white and peduncle flesh-colored. Beds of this species are known in shallow offshore waters in depths of less than 100 feet. The peduncle extends down into the soft sediments leaving the shells exposed above the surface. They feed by opening their two shells and filtering food from the water.

375

PHYLUM PHORONIDEA

The name Phoronidea is derived from Phoronis, a character in Greek mythology. This small group of animals are worm-like in appearance (Fig. 376). They have many tentacles at the anterior end and live in a tightly constructed sandy tube. At first glance they resemble polychaetes (for example, see Fig. 71), but the phoronids lack the body segmentation of the polychaetes. Two genera occur in local waters; the genus *Phoronis* lacks a collar at the base of their tentacles (Fig. 376) and the genus *Phoronopsis* (not figured) which possesses a collar.

376

Phoronis vancouverensis (Fig. 376). This species forms gray colored tubes which measure about 0.15 inches in diameter and 4-8 inches in length. The animal is white and is about 2 inches in length. Occasional specimens have been collected from the subtidal bottoms of Marina del Rey, Los Angeles Harbor, Alamitos Bay and Newport Bay. Tube masses of Phoronis vancouverensis have been collected from floating docks in Alamitos Bay and offshore pilings at Seal Beach.

PHYLUM CHAETOGNATHA
The arrow worms

The word Chaetognatha means bristle jaws, a characteristic feature of the arrow worms (Fig. 377). The bristles are found at the anterior end of the worm and surround the mouth. They are used to capture small animals and pass them to the mouth. The arrow worms, so named because of their general body shape, are transparent and swim in the water.

Sagitta euneritica (Fig. 377). This small species of arrow worm will measure up to 1 inch in length. It can only be collected with a plankton net in bays. Specimens of Sagitta euneritica have been collected in the offshore waters in Los Angeles-Long Beach outer harbor area and near the entrance of Alamitos Bay.

377

PHYLUM ECHIMODERMATA
The starfish, sea urchins, brittle stars, sea cucumbers

The echinoderms, or spiny skinned animals, are characterized by having calcareous spines which generally project out from the surface of the body. The symmetry of the animal is usually in fives or multiples of fives. All species of echinoderms live in the ocean. The common local echinoderms include the starfish (or seastars) (Figs. 378-387), the sea urchins and sand dollars (Figs. 388-392), the brittle stars (Figs. 393-400) and the sea cucumbers (Figs. 401-403).

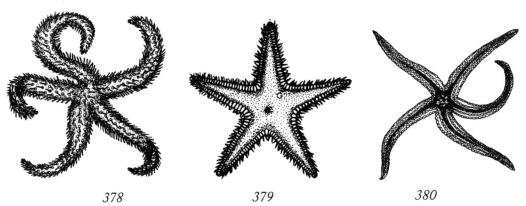

378   379   380

CLASS ASTEROIDEA—the starfish or sea stars (Figs. 378-387). The asteroids are star-shaped animals which typically have 5 rays or arms, or more in some species, which radiate out from a central disc. Short calcareous spines are present especially on the upper surface which form a characteristic pattern for each species. A groove is present along the undersurface of each arm where many soft, fleshy projections, the tube feet, are found. The tube feet function in locomotion and in food gathering.

Astrometris sertulifera—**the soft starfish** (Fig. 378). This species of starfish is so named because of greater than usual fleshy appearance. The soft starfish will measure nearly a foot in diameter. The background color is dark reddish brown with long orange and blue spines and yellow tube feet. The soft starfish is a predator feeding upon such animals as chitons, snails, barnacles, clams and brittle stars. Specimens of *Astrometris sertulifera* are found at rocky shores at low tides such as Flatrock, Pt. Fermin, Laguna Beach and Dana Pt.

Astropecten armatus—**the southern sand star** (Fig. 379). This starfish is gray in color and extremely large specimens attain diameters of 10 inches. Easily identified by the row of spines along the margin of the arms. The southern sand star feeds upon snails, sea pansies, polychaetes, tusk shells and dead animals. *Astropecten armatus* is more typically seen in subtidal offshore waters, but occasional specimens have been observed at low tides, or washed ashore, along the sandy shores of Long Beach and Laguna Beach.

Linckia columbiae (Fig. 380). This species measures 4-5 inches in diameter; it is generally gray in color with closely adhering scales on the arms. *Linckia columbiae* has strong powers of regeneration and specimens can often be seen with 1-5 arms in various states of growth. Specimens have been seen in minus tides associated with rocks at Lunada Bay and Flatrock Cove in the Palos Verdes Peninsula.

Lucida foliolata (Fig. 381) A large 5-rayed starfish which reaches 15 inches in diameter. The upper surface is gray in color and contains many short spines; the lower surface is

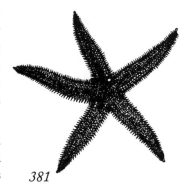

381

130

yellow. This species is more typical in subtidal offshore waters but an occasional specimen may be collected at rocky shores during minus tides.

Patiria miniata—the sea bat (Fig. 382). The sea bat, so named because of the apparent webbing between the short, thick arms, measures up to 6 inches in diameter. Its color is variable; typically it is red on the upper surface an yellow on the lower surface. However, the upper surface may be dark red, purple, greenish or mottled. The polychaete worm, *Ophiodromus pugettensis* (Fig. 48) is frequently found along the grooves on the undersurface of the sea bat. *Patiria miniata* has been seen at all rocky shores in the Los Angeles-Orange Counties area as well as the rock jetties at the entrance of Alamitos Bay and Newport Bay. Usually it is found under rocks at minus tides.

Pisaster giganteus (Fig. 383). This large starfish may measure up to 18 inches in diameter. The spines are more or less randomly distributed over the upper surface; they are large, bluntly-shaped, and set off from the orange-yellow colored surface. It feeds upon a variety of animals such as mussels, clams, snails and chitons. This starfish is only found occasionally on the rocks at minus tide level at Flatrock, White's Pt., Pt. Fermin, the jetty of Alamitos Bay, Little Corona and Laguna Beach.

Pisaster ochraceus—the ochre starfish or common starfish (Fig. 384). The common name of ochre starfish comes from one of its color phases-orange. This starfish may also be yellow, brown or purple in color. Specimens in Southern California may reach one foot in diameter. The spines form a pentagonal network on the upper surface of the central disc. The ochre starfish is the most commonly encountered

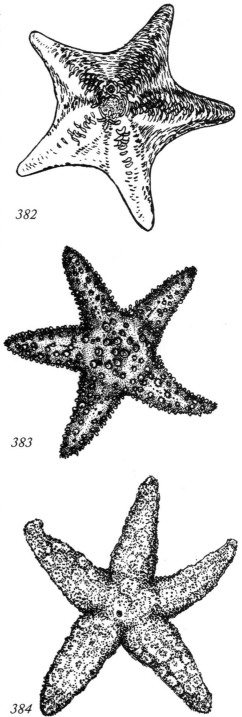

382

383

384

131

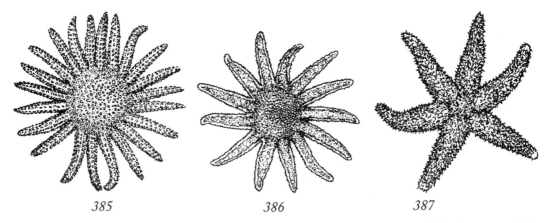

385   386   387

starfish in Southern California. However, it is becoming increasing difficult to see this species and other starfish because people take the specimens from the ocean. Specimens are seen at low tide at all rocky shores and many of the rock jetties where it feeds upon mussels. They are often seen clustered together along ledges just beneath beds of the California mussel, the principal diet of this starfish.

Pycnopodia heliantoides—**the sunflower star or the twenty-ray star** (Fig. 385). This is the largest species of starfish along the Paciffic coast; specimens will reach 2 feet in diameter. It is easily identified by its 20 or more rays. It is an actively moving starfish and feeds on sea urchins, crabs, chitons, other starfish as well as dead animals. It is rare in Southern California where is occurs along rocky shores at subtidal depths.

Solaster dawsoni—**Dawson's sun star** (Fig. 386). This species does not occur in Southern California but it is included since it is another starfish with many rays (8-13). It occurs on low intertidal rocky shores in northern California and northward.

Leptasterias hexactis (Fig. 387). This starfish measures 4-5 inches in diameter. It is usually black or brown in color but may be red or green. It is generally found among beds of California mussels where it feeds upon the different small animals present in this niche.

CLASS ECHINOIDEA—**the sea urchins and sand dollars** (Figs. 388-392). These animals are characterized by having movable spines. Echinoids may be globose in shape (Figs. 388-391) or discoidal in shape (Fig. 392).

Lovenia cordiformis—**the sea porcupine or heart urchin** (Fig. 388). The common names of this sea urchin come from the general outline of the body and its fewer but longer spines suggests a porcupine, and the animal has a heart-shaped outline when viewed from above. The sea porcupine burrows just beneath the sand in subtidal waters. Specimens have been taken from the channel entrance of Newport Bay and off Little Corona.

388

Strongylocentrotus franciscanus—the giant red sea urchin (Fig. 389). This is the largest of the local sea urchins; specimens measure up to 7 inches in diameter. This animal is brick red to a dark red or reddish in color. The spines are relatively longer in the giant red sea urchin than in the purple sea urchin. Both these species generally eat sea weeds, but they will eat debris and other organic matter. The giant red sea urchin is known from the low tidal zone of all rocky shores in Southern California and along the outer reaches of rock jetties leading into bays.

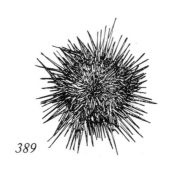
389

Strongylocentrotus purpuratus—the purple sea urchin (Fig. 390). This species is purple in color and measures up to 3 inches in diameter. The spines are shorter than the red urchin. The purple sea urchin is a characteristic animal of the low tide zone of all rocky shores where countless numbers of them may occur in tide pools, in burrows, broad shelves or under rocks. Specimens are found along breakwaters near the entrances of bays and harbors.

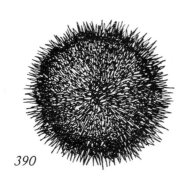
390

Lytechinus anameus (Fig. 391). The color of this sea urchin is yellow to gray with purplish spots on the upper surface. The body of the sea urchin is about 1.5 inches in diameter and the spine 1 inch in length. This species is not found intertidally, but is seen in shallow subtidal waters often in association with eel grass at the entrances of bays such as Newport Bay. This species has also been known under the name of *Lytechinus pictus*.

391

Dendraster excentricus—the sand dollar (Fig. 392). This gray colored sand dollar is discoidal in shape and measures about 3-4 inches in diameter. The flower-like pattern on the upper surface can be seen when the short spines are removed. *Dendraster excentricus* lives just beneath the surface of the sand; when removed, it is capable of burying itself rapidly. Patches of the sand dollar have been observed intertidally just within the entrance of Alamitos Bay and Newport Bay in sandy beaches; typically it occurs subtidally in offshore waters. Dead specimens, or pieces of specimens, are often seen washed up on sandy beaches.

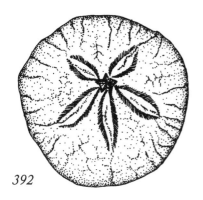
392

133

CLASS OPHIUROIDEA—the brittle stars or the serpent stars (Figs. 393-400). The common names for the ophiuroids are derived from the brittle nature of the arms and the ability of the arms to move rapidly and in a snake-like manner. Variations of the Greek root Ophio are used in many of the scientific names of brittle stars which means snake. The arms are distinctly set off from the central disc. They feed upon the fine particulate matter which collects between sand grains.

Amphiodia occidentalis (Fig. 393). This species of brittle star measures about 3-4 inches from one arm to an opposite one; it is characterized by 5 pairs of distinct radial shields on the central disc (Fig. 393) and its long slender arms with short spines. The disc is usually gray with some reddish markings and the arms white to yellow. Specimens have been collected during low tides from the sand under rocks at Little Corona and Laguna Beach. It also occurs subtidally in sandy sediments in which it can burrow.

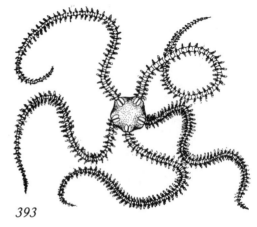

393

Amphipholis squamata (Fig. 394). The central disc measures about 0.3 inches in diameter and the entire animal about 2 inches across. The disc is distinctly pentagonal in shape with 5 pairs of narrow radial shields. This brittle star has been collected from under rocks at Flatrock and White's Pt. in the Palos Verdes area.

Ophioderma panamensis (Fig. 395). This brittle star is the largest of our local species; the disc may measure up to one inch in diameter and an arm spread of up to 7-8 inches. The spine of the arms are short and directed downward. This species is reddish-brown in color and is sometimes confused with *Ophioplocus esmarki* (Fig. 396). *Ophioderma panamensis* is found at low tide at such rocky beaches as White's Pt., Pt. Fermin, Laguna Beach and Dana Pt.

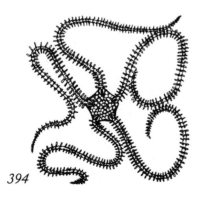

394

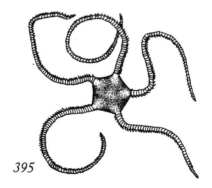

395

Ophioplocus esmarki (Fig. 396). This species measures about 3 inches across. It is pale brown in color. The arms resemble those of *Ophioderma panamensis* (Fig. 395). This brittle star is fairly common under rocks at several of the Palos Verdes shores and also it has been seen at Little Corona.

Ophiopterus papillosa (Fig. 397). Three common local species are characterized by having long spines on their arms (Fig. 397-399). This species lacks spines on the upper surface of the disc as is the case with the other 2 species. The disc measures less than 0.5 inches in diameter and is fleshy. Specimens measure about 2-3 inches across the arms. *Ophiopterus papillosa* is brown in color and has been found under rocks and protected niches at the rocky shores of Palos Verdes Peninsula, the jetty of Alamitos Bay, Little Corona, Laguna Beach and Dana Pt.

Ophiothrix rudis (Fig. 398). This brittle star can be distinguished from *Ophiothrix spiculata* (Fig. 399) with the aid of a hand lens by having 5-6 smooth spines of each side of an arm segment as compared to 7 serrated ones. *Ophiothrix rudis* is green to tan in color with reddish bands of the arms. It has been taken at low tides from under rocks at White's Pt., Pt. Fermin and Dana Pt.

Ophiothrix spiculata—the spiny brittle star (Fig. 399). This species is gray in color and may have orange spots at the base of each arm. It will measure up to 6 inches across the arms. It has been taken in protected niches at low tides along the rock jetty of Alamitos Bay and Newport Bay and from under rocks at Little Corona and Laguna Beach.

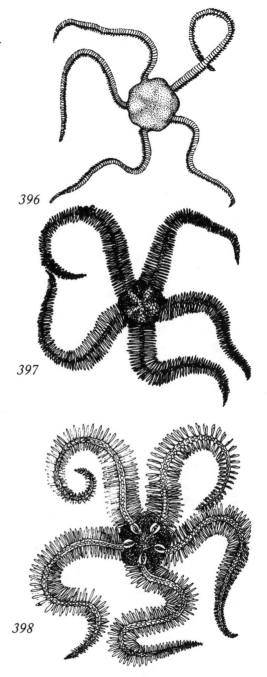

396

397

398

Ophionereis annulata—the banded brittle star (Fig. 400). This species is so named because of the segmented occurrence of black pigment along the gray arms which makes it easy to

399

400

identify. The banded brittle star measures 4 inches across the arms. It has been seen at many localities at low tides under rocks such as Flatrock, White's Pt., Pt. Fermin, Newport Bay entrance, Little Corona, Laguna Beach and Dana Pt.

CLASS HOLOTHUROIDEA—the sea cucumbers (Figs. 401-403). The sea cucumbers are not typically appearing echinoderms. The surface of the body lacks spines; they are embedded in the tissue, and a microscope is required to see them. The body surface may be wart-like (Fig. 403) or smooth (Figs. 401, 402). The sea cucumbers have a mouth at one end which is surrounded with circle of tentacles(Fig. 326) and an anus at the opposite end. The sea cucumbers feed upon organic matter present with sand and debris.

Leptosynapta albicans (Fig. 401). This small white colored sea cucumber measures about 2-3 inches in length. The body is smooth and lacks tube feet. It burrows into sand or mud and it can be collected intertidally at low tides from the sandy mud beaches of Alamitos Bay, Anaheim Bay and Newport Bay.

401

Caudina arenicola—the sweet potatoe cucumber (Fig. 402). The common name comes from its general body shape and also from similar color to the sweet potatoe. It measures up to 5-6 inches in length. It burrows into the finer sediments to depths of 1.5 feet at such beaches as Long Beach and Seal Beach.

402

403

Parastichopus parvimensis—the **Southern California** sea cucumber (Fig. 403). This sea cucumber is red-brown in color and will measure 12-18 inches in length. The surface has many papillae; the tube feet occur as 5 longitudinal rows on the undersurface. It feeds by taking in sediment and digesting the organic matter present in much the same way as an earthworm. The Southern California sea cucumber is typically a subtidal inhabitant of rocky shores but occasional specimens have been seen on the rock jetty of Alamitos Bay and the shores of Little Corona, Laguna Beach and Dana Pt. at minus tides.

PHYLUM CHORDATA
The tunicates, fishes, amphibians, reptiles, birds, and mammals

The chordates (Figs. 404-521) are characterized by the possession, at least during some stage of their life, of gill slits, a dorsal, hollow, central nervous system and a notochord (this structure is replaced by a backbone in the vertebrates). While the best known and most common species of chordates are the vertebrate animals (fishes, amphibians, reptiles, birds and mammals), there are 3 groups of protochordates which are forerunners of the vertebrates; these are the Urochordata, or tunicates (Figs. 404-416), the Cephalochordata (Fig. 416) and the Hemichordata (Fig. 418).

SUBPHYLUM UROCHORDATA—the tunicates or sea squirts (Figs. 404-416). The urochordates or tunicates are entirely a marine group of animals which may be solitary (Figs. 406, 412-414), colonial (Figs. 404, 405, 407-410) or pelagic (Figs. 415, 416). The notochord and central nervous system are present only during the larval stages; the gill slits persist as the pharynx which is the food gathering structure. Tunicates feed by pumping water through their bodies, hence the common name sea squirts, and removing the fine particulate matter from the water (the pharynx is located on the left side of Fig. 406). The colonial species secrete a matrix of cellulose which may completely hide the individuals of the colony (as in Fig. 404). The individuals of the colonial species appear much like the solitary forms, but much smaller, such as shown in Figure 406.

Amaroucium californica—the sea pork (Fig. 404). The common name is derived from the superficial appearance of this animal to salt pork. Most of the colony is composed of the cellulose matrix and a dissecting microscope is necessary in order to see the individual members of the colony. The individual colonies may reach 1-3 inches in diameter. The sea pork is a common inhabitant of the local marinas, especially during the late summer months; it has been found growing over the surface of the bay mussels and other animals at Marina del Rey, Alamitos Bay, Huntington Harbour and Newport Bay.

404

Botryllus sp. (Fig. 405). This colonial species consists of many individuals units, one of which is shown in Figure 405. Each unit is composed of several individuals, each of which has a single incurrent siphon (dark circles in Fig. 405) for water intake, and they all share a common, centrally located excurrent siphon for water discharge. The colony is thin but it may cover an area of several inches. It is orange-red to purple in color. Specimens have been found on floating boat docks especially growing over the bay mussels in Los Angeles outer harbor, Alamitos Bay and Newport Bay.

405

Ciona intestinalis (Fig. 406). This solitary species of tunicate is soft to touch and in expanded state it will reach 3-6 inches in length. The color is generally pale yellow to green but in larger, older animals the surface may be covered with debris or other animals. *Ciona intestinalis* may literally cover the under surface of boats or boat floats during the spring or fall months. Growth of the animal is rapid; sexual maturity may be reached within 2 months. The sexes are separate in this species, and it is a convenient source of material to demonstrate fertilization and early cleavage in the laboratory. It has been found in all bays, harbors and marinas in Southern California attached to boat floats, boats and pilings.

406

Diplosoma pizoni (Fig. 407). This small, flat colonial tunicate has individual animals scattered randomly in a common gelatinous matrix. Colonies will reach about 1 inch in diameter. They are yellow and brown in color. They have been found in the outer portions of Los Angeles-Long Beach Harbors, Alamitos Bay and Newport Bay where they are found growing over the surfaces of the bay mussel or the solitary tunicate *Styela plicata* (Fig. 414) which are attached to floating boat docks.

Eudistoma diaphanes (Fig. 408). This species of colonial tunicate is transparent to pale yellow in color. Specimens will measure up to 2 inches in diameter. It has been collected from the surface of rocks at low tides at Newport Bay and Little Corona.

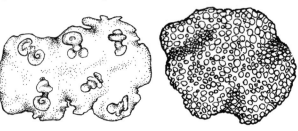

407   408

Euherdmania claviformis (Fig. 409). The individuals of the colonies are finger-shaped and pale green in color. They will measure about 1.5-2 inches in length. Clusters of *Euherdmania claviformis* may be found along the rocky shores of Palos Verdes

138

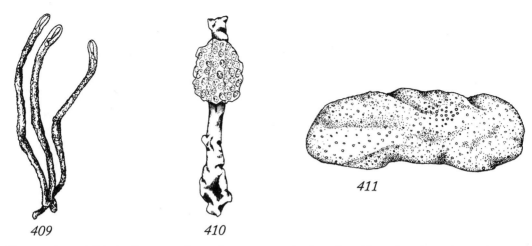

409 *410*

411

Peninsula and Little Corona where they are attached to the sides of rocks at low tide.

Perophora annectens (Fig. 410). This pale yellow colonial tunicate is composed of many individuals which are found in a transparent gelatinous matrix. The colonies typically attach to algae (Fig. 410) or sides of rocks at low tides at the rocky shores of Southern California.

Polyclinum planum (Fig. 411). Colonies of the gelatinous appearing tunicate will reach 4 inches in length. It is tan to light brown in color. It is found attached to the undersides of rocks in the intertidal shores of Pt. Fermin and the jetty of Newport Bay. It is known to occur on the upper surface of rocks subtidally.

Pyura haustor (Fig. 412). A solitary tunicate which is irregular is shape but with well defined incurrent and excurrent siphons. Individuals are variously colored depending upon what other organisms are encrusting on its surface, but a clean specimen is orange to red. Specimens will measure up to 2 inches in height. It has been found on pilings in Newport Bay.

412

Styela montereyensis (Fig. 413). This long, solitary tunicate measures 6-10 inches in length. It is dark red in color, but the color may be altered because of the presence of other organisms attaching to its surface. *Styela montereyensis* often forms dense colonies on pilings in the protected waters of Alamitos Bay, Anaheim Bay, Huntington Harbour and Newport Bay. Specimens will attach to the lower surfaces of boat floats.

413

Styela plicata (Fig. 414). This solitary tunicate is white in color and possesses many wrinkles giving it an appearance of a brain. Large specimens will measure up to 2 inches in diameter. At times the incurrent an excurrent siphons may be difficult to locate because of the wrinkles. *Styela plicata* is found attached to pilings and boat floats in Alamitos Bay, Huntington Harbour and Newport Bay. It is more commonly encountered in the spring and fall months.

Pyrosoma giganteum (Fig. 415). This cylindrical shaped pelagic animal will measure up to 1 foot in length. The body wall contains many cone-shaped protuberances. It is bioluminescent at night when disturbed. It is uncommon along the Southern California coast, but it is often seen at night while anchored at Catalina Island.

Salpa sp. (Fig. 416). This pelagic tunicate occurs singly or in long chains. It is usually seen as chains which may be a few feet in length. All individuals of the chain are the same age. This species is usually seen in the spring in the channel between the mainland and Catalina Island.

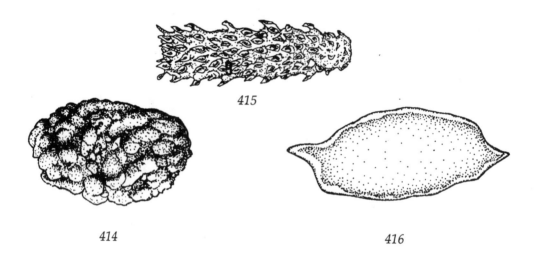

415

414

416

SUBPHYLUM CEPHALOCHORDATA—the lancelets (Fig. 417). The notochord, gill slits, dorsal nerve cord are present in both the larval and adult stages.

Branchiostoma californiense (Fig. 417). This species of lancelet is flesh colored and will measure 1-4 inches in length. It is fish-like in appearance but lives buried in the sand. It has

been collected subtidally at the entrance to Marina del Rey, Alamitos Bay and Newport Bay. Specimens have been collected at the entrance of Newport Bay during minus tides. Specimens have been reported from the intertidal sandy beach at Little Corona.

417

140

SUBPHYLUM HEMICHORDATA (Fig. 418). This group is worm-like animals is sometimes referred to as a separate phylum. They are entirely marine and are easily distinguished by an anterior proboscis which is set off from the rest of the animal. A useful diagnostic feature of this group is the presence of an iodoform odor in living specimens. They live in sand or mud a few inches beneath the surface and feed upon the organic material present in the substrate.

Saccoglossus **sp.** (Fig. 418). This species of hemichordate is the only intertidal representative of the group known from the local area. It is yellow-orange in color and measures 2-3 inches in length. It has only been found from the intertidal beaches of Balboa Island in Newport Bay.

418

SUBPHYLUM VERTEBRATA—the vertebrates (Figs. 342-395). This group includes the fish, amphibians, reptiles, birds, and mammals; the fish, birds, and mammals are important marine inhabitants and representatives of these groups will be discussed herein. The backbone is a characteristic feature of the vertebrates; the backbone consists of many vertebrae which are composed of cartilage or of bone.

CLASS CHONDRICHTHYS—sharks and rays (Figs. 419-424). The sharks and rays differ from the true fish in that their skeleton is composed of cartilage rather than bone. The mouth opens on the lower surface (Fig. 419) rather than at the end (Fig. 432, for example) as in the true bony fish. Many additional species of sharks and rays are known to occur in Southern California but only a representative group are discussed herein.

Family Carcharhiidae (Figs. 419, 420). This family is known as the gray sharks. Its first dorsal fin is located in front of the ventral fin and the second dorsal fin is opposite the anal fins.

Triakis semifasciata—**leopard shark** (Fig. 419). The leopard shark is easily distinguished from other species by its black spots scattered over its body. It occurs from Oregon to Mexico including the Gulf of California. Specimens will reach 6.5 feet in length. It occurs

419

in many of the bays of Southern California and in shallow offshore waters. The food habits of the leopard shark vary with age. Younger specimens feed on worms and small crustaceans; as they grow older they feed on shrimp and crabs as well as fish.

Mustelus californicus—**gray smoothhound shark** (Fig. 420). This species is a common shark in Southern California. The skin is smooth and the upper surface gray and whitish below. The gray smoothhound shark will measure up to 5 feet in length. It occurs in all bays and harbors of Southern California and shallow offshore waters. It is known from northern California to Mazatlan, Mexico.

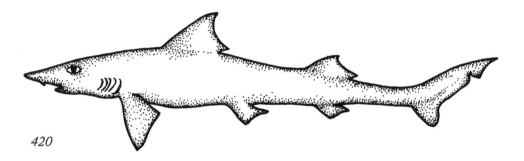

420

Family Rhinobatidae—The guitarfishes (Fig. 421). The common name comes from their general body shape to that of a guitar. The guitarfish are rays which resemble sharks in part since their head is elongated and the posterior part of the body is well developed; however, the gill slits are located under the pectoral fins (the anterior pair of fins) which are characteristic of rays.

Rhinobatos productus—**the shovelnose guitarfish** (Fig. 421). This species is gray in color on the upper surface and whitish on the lower surface of the body. Specimens will reach 4 feet in length. The snout is sharply pointed. The shovelnose guitarfish lives in the shallow water along the coast and in bays from Monterey south into Baja California. It feeds upon bottom-dwelling crustaceans and clams.

Family Rajidae (Fig. 422). There are short spines located along the mid-body extending into the slender tail. Eggs are laid in large, dark brown leathery cases which are more-or-less square in shape with long spines extending out from each corner.

Raja inornata—**California skate** (Fig. 422). The snout is pointed. It is olive brown on the upper surface and tan below. Spines are present on the lateral "wings" and tail. It has been reported from Puget Sound to Baja California. Specimens measure up to 2.5 feet in length.

Family Myliobatidae (Fig. 423). The head projects forward and is distinct from the main body. The tail is long and narrow and lacks fins but has one or more venomous spines.

142

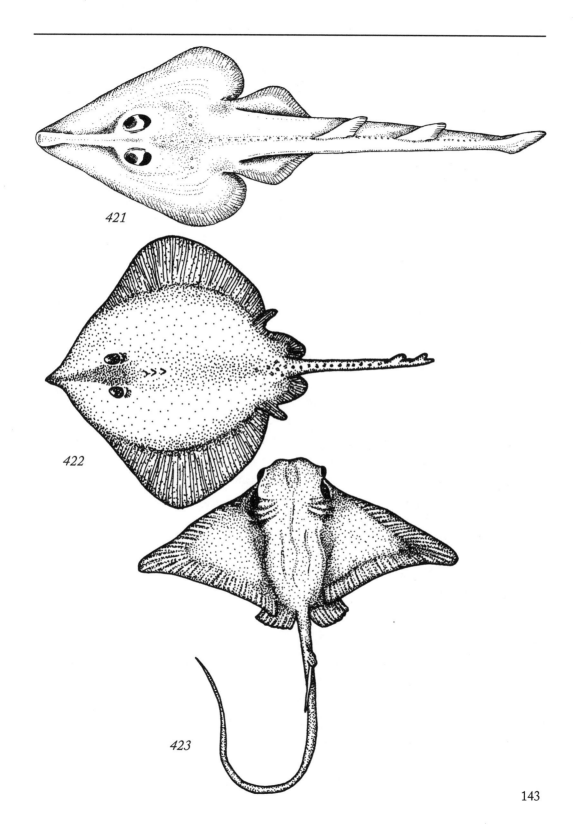

421

422

423

Myliobatis californica—Bat ray (Fig. 423). It is darkly colored on the upper surface and lighter below. A poisonous spine is present on the dorsal surface of the tail near where it extends from the body. Specimens attain lengths of 4 feet. It is known from Oregon to Baja California.

Family Dasyatidae—the stingrays (Fig. 424). The pectoral fins are more or less rounded and considerably larger than the pelvic fins. Dorsal fins are absent, but a tail fin is present. A spine is found along the upper surface of the tail which contains a toxic poison. The sting from the local species is painful and if a person is stung, which is usually caused by stepping on this bottom-dwelling stingray, bleeding should be encouraged and a physician consulted. Annual drives to control the stingrays are conducted in Southern California for example, at Seal Beach. The stinger can be cut off and the stingray released.

Urobatus halleri—**the round stingray** (Fig. 424). The upper surface of this species is brown and frequently spotted; it is yellow underneath. The round stingray will reach 18-20 inches in length. It feeds upon bottom-dwelling animals such as worms, clams and fish. This species is the commonly encountered stingray especially in Alamitos Bay, the mouth of the San Gabriel River, Anaheim Bay and Newport Bay; it is also found in shallow waters of Southern California.

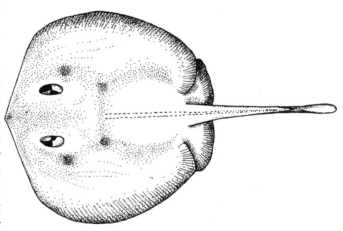

424

CLASS OSTEOICHTHYES—the bony fishes (Figs. 425-480). The bony fish or true fish differ from the sharks and rays in that their skeleton is composed of bone and their mouth is located at the terminal end of the head. Since many fish are protected by laws as to season, catch, gear, etc., the person should consult the Department of Fish and Game as to regulations.

Family Muraenidae—the moray family (Fig. 425). The moray family can be distinguished by its eel-shaped body and the absence of pectoral fins.

Gymnothorax mordax—**the California moray or moray eel** (Fig. 425). The only member of this family found in California waters. The dorsal and anal fins lack rays; pectoral and pelvic fins absent. Teeth numerous and very sharp. Color of body is mottled brown and green. It will grow to about 5 feet in length. More frequently present in subtidal water along rocky shores, but it can be seen in rocky crevices at minus tides at any of the local rocky shores. The California moray can attack and inflict a wound on a person's hand. It is esteemed as food by some people.

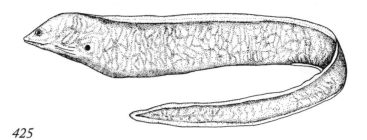

425

Family Clupeidae (Fig. 426). This family contains the herrings, sardines and shads which makes this a commercially important family. The lower jaw may extend beyond the upper jaw. The head lacks scales.

Sardinops sagax caeraleus—the **Pacific sardine** (Fig. 426). It is blue-green in color above and white below. It will measure 16 inches in length. The Pacific sardine is known from

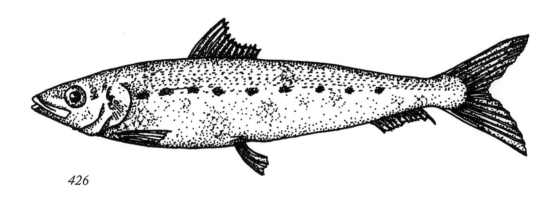

426

Alaska to Guaymas, Mexico. The Pacific sardine was at one time the largest commercial fisheries in United States. Fish harbor in Los Angeles Harbor was an important locality for sardine canneries as well as other areas of California including the famous cannery row of Monterey. The peak in cannery operations occurred just after World War II and collapsed in the 1950s. The cause of the demise of the sardine populations is unknown, but it was undoubtedly related directly or indirectly to overfishing.

Family Engraulidae—the anchovies (Fig. 427, 428). The distinctive characteristics of the anchovy family include the lack of scales on the head and the forward projecting upper jaw with reference to the lower jaw. A single dorsal fin located near the middle of the body is also present.

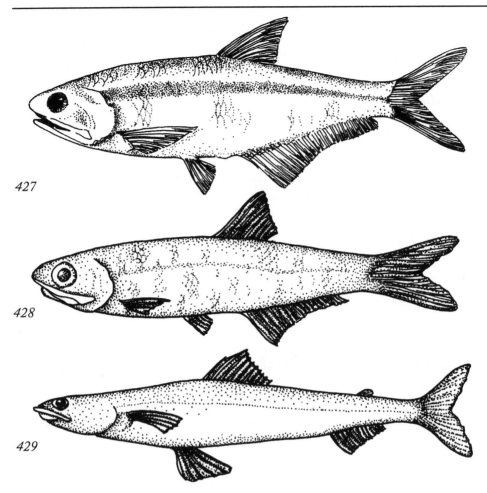

427

428

429

Anchoa compressa—**the deepbody anchovy** (Fig. 427). A tan colored fish with a light band extending along the sides of the body. This species of anchovy measures about 6 inches in length. The anal fin is long and nearly reaches the base of the tail fin. The deepbody anchovy is found in schools from Central California south into Baja California both offshore and within the local bays and harbors.

Anchoa delicatissima—**slough anchovy** (Fig. 428). It is brown to greenish in color on the upper surface and silvery below. Specimens measure less than 4 inches in length. It is present from Southern California to Baja California.

Family Synodontidae—**the lizard fish** (Fig. 429). Jaws long extending posterior to the eyes. The teeth are large, canine-like.

Synodus lucioceps—**the California lizard fish** (Fig. 429). Color is brown above to tan below. It will measure 25 inches in length. It is distributed from San Francisco to Guaymas, Mexico.

146

Family Batrachoididae—the toadfish and midshipmen (Fig. 430). The toadfish are bottom-dwelling species with large mouths. The midshipmen possess numerous light producing organs, called photophores, which are arranged in rows along the head and sides of the body. They appear as shiny white spots.

Porichthys myriaster—**the slim midshipman** (Fig. 430). The ground color is dark bronze with purple to blue tinges. The color is generally lighter below and may become a golden yellow. This species grows to lengths of 15 inches. *Porichthys myriaster* is known from Southern California to Baja California in shallow waters and in bays such as Alamitos Bay, Anaheim Bay

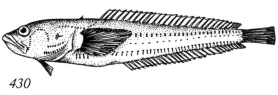

430

and Newport Bay. This species is not important as either a commercial or sport fish.

Family Gobiesocidae—the clingfishes (Fig. 431). These fish resemble gobies in that the 2 pelvic fins join to form a sucker. They differ in that the clingfish have a single dorsal fin and the gobies have two.

Gobiesox rhessodon (Fig. 431). This small clingfish measures a few inches in length. It is brown in color with darker brown spots scattered over its surface. This species occurs in

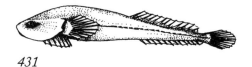

431

both bays and offshore waters. It can be seen clinging to the undersurface of larger rocks at many local rocky shores at low tides. It is known from Southern California to Baja California.

Family Cyprinodontidae—the killifish (Fig. 432). One dorsal fin and the other fins broadly rounded.

Fundulus parvipinnis—**the California killifish** (Fig. 432). Small fish measuring up to 4 inches in length. It is widely distributed from San Francisco Bay to Peru. It feeds on small

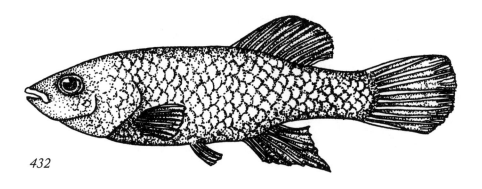

432

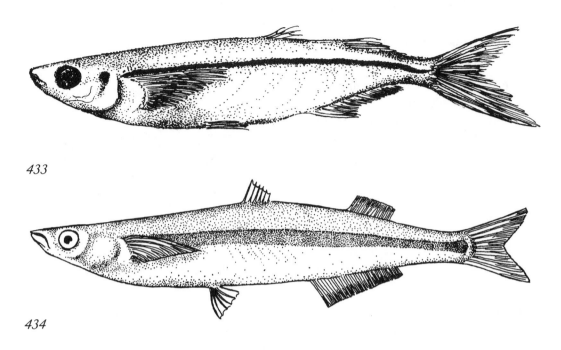

433

434

crustaceans and insects. The California killifish lives in estuarine to nearly fresh water in the upper tidal reaches of Ballona Creek, Los Angeles River, San Gabriel River, Santa Ana River as well as the back bay areas of Alamitos and Newport Bays.

Family Atherinidae—the silverside fish (Figs. 433-435). This family includes the jacksmelt, the topsmelt and the grunion. They have 2 separate dorsal fins, the first of which has spines and the second of which has soft rays. A silver streak runs along the side of the body which accounts for the common name of the family.

Atherinops affinis—**the topsmelt** (Fig. 433). This species is separated from the other members of the family by having forked teeth. The topsmelt measures up to 12 inches in length. It is blue-gray in color on the upper surface and becoming silver on the lower surface; the silver streak is found on either side extending from the region of the gills to the tail fin (Fig. 433). They inhabit inshore waters of Southern California, often in schools, and extend into such protected waters as Alamitos Bay, Lower San Gabriel River, Anaheim Bay and Newport Bay. They extend from Oregon into the Gulf of California.

Atherinops californiensis—**the jacksmelt** (Fig. 434). This species has simple teeth in contrast to the forked teeth of *Atherinops affinis* (Fig. 433). It is gray-green in color with the sides and lower surface metalic. This fish reaches upwards to 2 feet. The jacksmelt constitutes the most important smelt in California; it is also used as bait and is caught by fishermen from piers. It is found in inshore waters along the coast.

148

Leuresthes tenuis—the California grunion (Fig. 435). This fish lacks teeth which separates it from the other 2 species in the topsmelt family. The grunion is blue-green along the upper surface and silvery along the lower surface. A bright silver blue to violet band extends along the sides of the body. It grows to 7 inches. The grunion spawns at night during the highest tides from late February until early September on many Southern California sandy beaches. The eggs are buried in the sand where development follows until the next series of high tides which uncovers the eggs and the larvae hatch out and swim away. They may only be caught with the hands by sportsfishermen. Consult the Department of Fish and Game of California for current laws and regulations governing the California grunion.

Family Sygnathidae—the pipefish and seahorse family (Fig. 436). The mouth is present at the end of a tubular snout. The eggs are brooded by the male in a pouch under the tail (seahorses) or attached to the abdomen or tail which may or may not be protected with pouch flaps (pipefish).

Syngnathus griseolineatus—**the bay pipefish (Fig. 436).** This light brown, slender fish grows to nearly a foot in length. It possesses a single small dorsal fin and is known from bays and estuaries from Alaska to Southern California. It is often seen swimming around the pilings in the Marina.

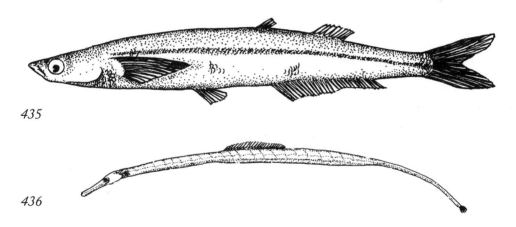

435

436

Family Scorpaenidae—the rockfish (Figs. 437-441). The rockfish are characterized by the possession of a bony support extending downward from the region between the eyes and gill slits(shown in Figs. 437-441). The common name for the family comes from its typical habitat of living along the rocky coasts. The rockfishes are important as a commercial and sport fish.

Sebastes goodei—**the chilipepper (Fig. 437).** The lower jaw projects beyond the upper jaw. It is pink in color with a pink line extending the length of the lateral line. The common

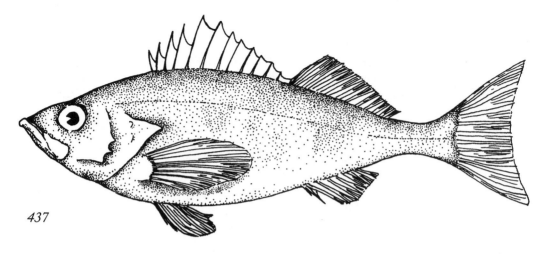

437

name comes from its color. It grows to about 2 feet in length. It is found all along the California coast south into Baja California in the vicinity of rocky shores.

Sebastes melanops—**blue rockfish** (Fig. 438). The adults are blue in color and the young are reddish. The blue rockfish reaches 20 inches in length. It is characterized by having spines in its dorsal fin and the anal fin is slanted or straight. This species is found from the Bering Sea to Baja California.

Sebastes paucispinis—**the bocaccio** (Fig. 439). It can be distinguished from the chilipepper by its olive to brown color with shading into orange along the sides to white on the

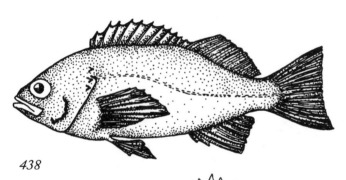

438

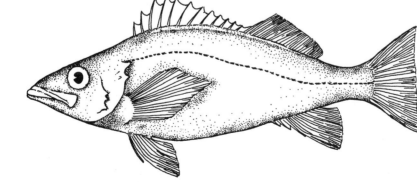

439

150

undersurface. It can grow to lengths of 3 feet and weigh up to 20 pounds. It is present from British Columbia to Baja California. The bocaccio is an important sport fish in Southern California especially off rocky shores.

Sebastes saxicola—stripedtail rockfish (Fig. 440). The common name is based on the presence of green stripes in the tail fin. It grows to 15 inches in length and is found in subtidal waters to depths of over 1000 feet. The stripedtail rockfish is distribed from southern Alaska to Baja California.

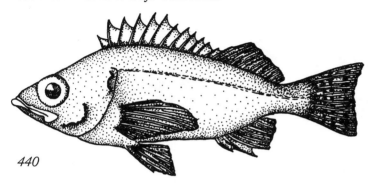

440

Scorpaena guttata—the sculpin (Fig. 441). This species is not a true sculpin (Figs. 442-445). The sculpin has several spines along the head and 12 sharp spines along the anterior half of the dorsal fin. Its upper surface has a backgroung color red which is mottled with various shades of color; it is generally pink on the undersurface. The sculpin is known from Central California south into Baja California. It is found along rocky shores and extends into the major local bays. The sculpin is important as a commercial and sport fish especially in Southern California.

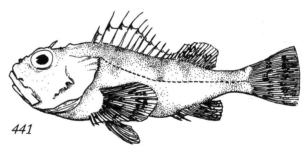

441

Family Cottidae—the sculpin family (Figs. 442-445). The members of this family have large eyes which are mounted on top of large heads. Sharp spines are located on the head and in back of the eyes. Some of the species of the family, including *Leptocottus armatus* (Fig. 444) and *Scorpaenichthys marmoratus* (Fig. 445), either lack scales or they are embedded in tissue.

Clinocottus analis—the wooly sculpin (Fig. 442). The wooly sculpin is so named because of the presence, especially in older specimens, of small spinous outgrowths from the surface of the head and parts of the body giving it a wooly appearance. This fish is grayish green

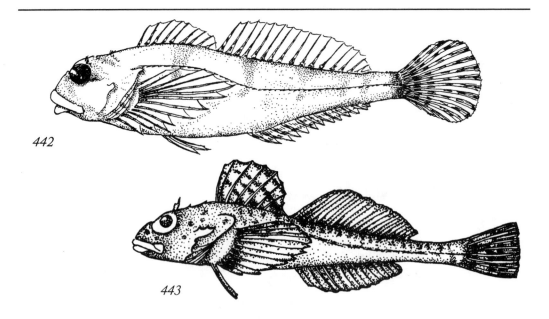

442

443

to olive brown in color with flecks of lighter spots over the body. Scales are present but they are deeply embedded in the tissue. This fish grows to several inches in length. It has been collected from Northern California to Baja California in shallow waters, especially in tide pools. It is an occasional inhabitant in bays.

Icelius quadriseriatus—**yellowchin sculpin** (Fig. 443). It is yellow on the underside of the head region; the color is dark brown to gray on the upper surface. The yellowchin sculpin is found from northern California to Cabo San Lucus, Baja California at depths of 20 to 300 feet.

Leptocottus armatus—**the staghorn sculpin** (Fig. 444). The staghorn sculpin is so named because of an antler-like spine in front of the operculum, or covering, of the gills. The upper surface is a mottled olive-gray-black color with the lower surface whitish. The dorsal

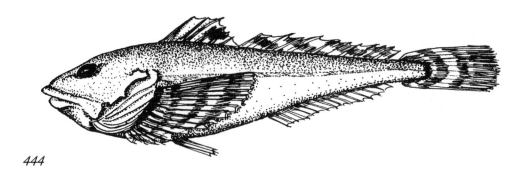

444

fin has dark pigmented spot towards the posterior end. This fish is known to grow to one foot in length. *Leptocottus armatus* occurs from Alaska to Baja California especially in inshore waters, bays and occasionally in fresh water. This fish is used by fishermen for bait.

Scorpaenichthys marmoratus—the cabezon (Fig. 445). This species may be recognized by the spine above each eye. The color of this fish is variable; the ground color is green and it is generally mottled from dark brown or gray to the tans or reds. The cabezon will grow to 30 inches in length and weigh up to 25 pounds. This species is known from British Columbia to Baja California especially in inshore waters and occasionally in bays. Crabs constitute an important element of their diet. The cabezon is an important game fish, but the eggs are poisonous and can cause severe illness.

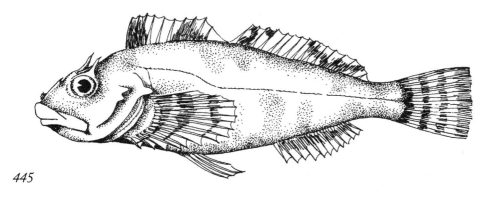

445

Family Serranidae—the bass family (Figs. 446, 447). The pelvic fins are located beneath the pectoral fins and consist of 5 rays. The lower jaw protrudes slightly. Three spines are present in front of the anal fin.

Paralabrax clathratus—the kelp bass (Fig. 446). A long dorsal fin has a sharp notch between the anterior spines and the soft posterior rays. The kelp bass grows to about 2 feet in length and is gray, brown to green in color with patches of lighter colors on the side and yellow markings on the fins. This species is an important sport fish and, as the common name suggests, it is found swimming among the fronds of the giant kelp *Macrocystis pyrifera*

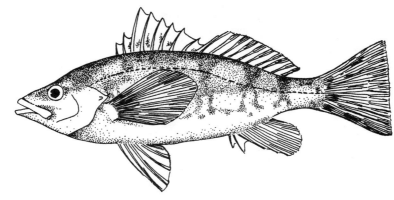

446

153

(Fig. 540) in Southern California. It is known from Central California south into Baja California.

Paralabrax nebulifer—the sand bass (Fig. 447).   The background color is gray with a greenish cast; the under parts are gray to white. The separation between the anterior spines and posterior soft rays of the dorsal fin is more gradual than in the kelp bass; in addition, the longer spines vary in length in the sand bass. The sand bass will grow to 20 inches in length. This species is known from Central California to Baja California. It is found in shallow offshore waters, but it will enter bays. It is a sport species and of minor importance as a commercial fish.

Family Pristipomatidae—the grunt family (Fig. 448).   The members of this group are referred to as the grunt family because of the sound they make by grinding their pharyngeal teeth together; the sound is intensified by the swim bladder. The family is characterized by very strongly developed pharyngeal teeth.

Anisotremus davidsoni—the California sargo (Fig. 448).   This fish is silver in color with a conspicuous dark vertical band behind the pectoral fins; additional smaller dark bands may be seen towards the tail. Specimens grow to 20 inches in length. The California sargo is known from Santa Barbara south into the Gulf of California. It occurs in offshore waters

447

448

154

as well as in bays, such as Alamitos Bay and Newport Bay. It constitutes a minor sport fish in Southern California.

Family Sciaenidae—the croaker fish (Fig. 449-451). This common name for this family comes from the grunt-like noise made by these fish. These sounds are made by rapid contractions of the strong muscles attached to the swim bladder; the swim bladder acts as a resonance chamber. The sound production by croakers increases during the spawning season, but the reason for this is unknown.

Chelotrema saturnum—**black croaker** (Fig. 449). The back of the fish, pelvic fins and the edge of the gill cover are black. It reaches 15 inches in length. It is found in intertidal waters to about 150 feet in depth. The black croaker is found from Point Concetion, California, south to Magdalena Bay in Baja California.

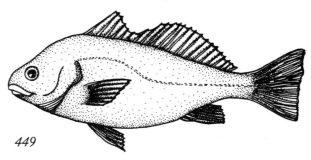

449

Genyonemus lineatus—**the white croaker** (Fig. 450). The white croaker is silver in color above and becoming white on the lower surface. Length ranges up to one foot. This species is known from British Columbia south to Baja California. In Southern California it is found in sandy inshore waters and within Alamitos Bay and Newport Bay.

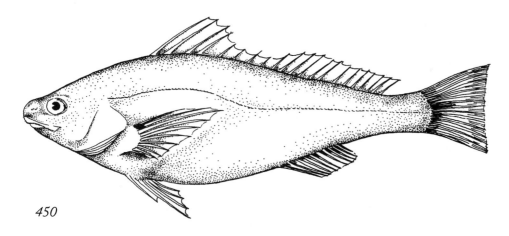

450

Menticirrhus undalatus—**the California corbina** (Fig. 451). The body is gray in color. The second anal fin spine extends beyond the margin of the fin. Specimens have been recorded to reach 28 inches in length. The California corbina extends from Point Conception, California south to Baja California.

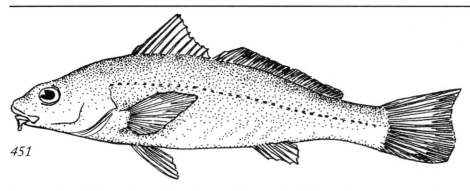

451

Family Girellidae—the nibbler family (Fig. 452).   These fish have hinged lips and many fine teeth with which they can nibble at all types of food very efficiently, hence the common name for the family.

Girella nigricans—the **opaleye** (Fig. 452).   The opaleye can be identified easily by the white spot on either side of the back. The common name, however, comes from the opalescent

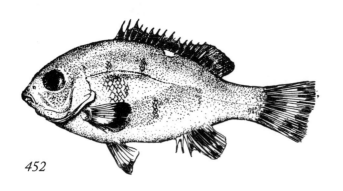

452

blue color of its eyes. The opaleye is greenish-blue in color with a lighter cast to the lower surfaces. It reaches about 20 inches in length. The dorsal fin contains 14 spines in front of soft rays. This species is known from Monterey south into Baja California. The young are found in tide pools along all the larger of the local bays. *Girella nigricans* is of importance as a surf sport fish.

Family Embiotocidae—the surfperches (Figs. 453-457).   The surfperches are separated from the true perches by having 3 anal spines rather than 2 in front of the anal fin. All species of surfperches bear their young alive; they generally swim into bays to give birth to their young.

Cymatogaster aggregata—the **shiner perch** (Fig. 453). This species attains lengths of up to 6 inches. It is silver in color with a greenish hue above; 3 yellow lines are present behind the pectoral fins and 8 dark lines can be seen along the lower

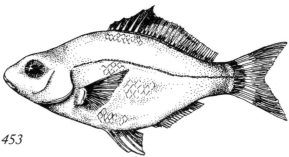

453

surface. The shiner perch ranges from Alaska to Baja California and can be found along sandy beaches and in bays.

Embiotoca jacksoni—the black perch (Fig. 454). The black perch varies in color; its background color is generally brown and tinged with blue, green, red or yellow. The pelvic and anal fins may be tinged with yellow or orange. This fish will grow to 14 inches in length. A patch of larger scales are present between the pectoral and pelvic fins. It is known from central California to Baja California from sandy beaches and bays. It is a minor sport fish; specimens are caught generally in bays around pilings.

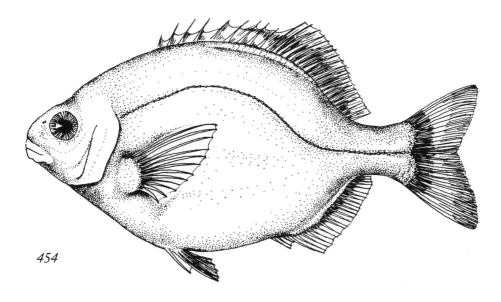

454

Hyperprosopon argenteum—the wall-eye surfperch (Fig. 455). The common name comes from its characteristic feature of having very large eyes. The fish is steel blue in color on the upper surface becoming white on the lower surface. The fish grows to one foot in length. The pelvic and caudal fins are tinged with black pigment at their margins. The wall-eye perch has been taken from British Columbia south in Baja California especially from sandy beaches such as along the Southern California coast. It is of importance in California both as a commercial fish and as a sport fish.

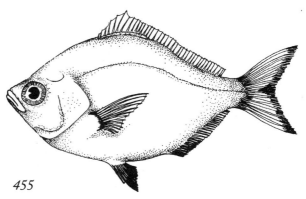

455

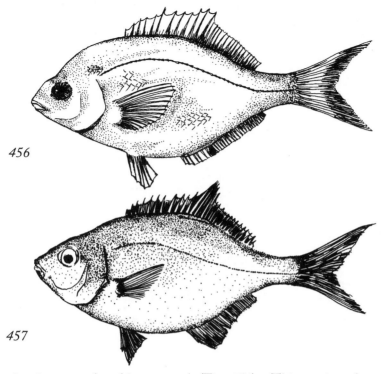

456

457

Phanerodon furcatus—the white seaperch (Fig. 456). This species of seaperch is white in color with a darker cast along the upper surface. Specimens reach lengths of one foot. Four to 5 rows of larger scales are present along the sides in the region between the dorsal fin and the lateral line. The white seaperch is found from British Columbia to Baja California especially along the sandy beaches; specimens sometimes enter bays such as Alamitos Bay and Newport Bay. This species is extremely important as a commercial fish in all coastal regions of California.

Rhacochilus vacca—the pile perch (Fig. 457). The ground color of the pile perch is silvery with a black cast on the upper surface and lighter below. The fins tend to be dark in color. This fish reaches lengths of 16 inches. The tail fin is deeply notched. There are 6 or more rows of fish scales between the dorsal fin and lateral line. *Rhacochilus vacca* ranges from Alaska to Baja California along sandy beaches, around kelp and around pilings. It is of importance as a commercial fish.

Family Pomacentridae—the damselfish (Figs. 458, 459). A largely tropical marine family of fish which the majority are small in size. Most of them are brilliantly colored such as the garibaldi (Fig. 459).

Chromis punctipinnis—blacksmith (Fig. 458). Black spots on posterior half of the fish including anal fin and on the dorsal fin. It is common in Southern California especially in

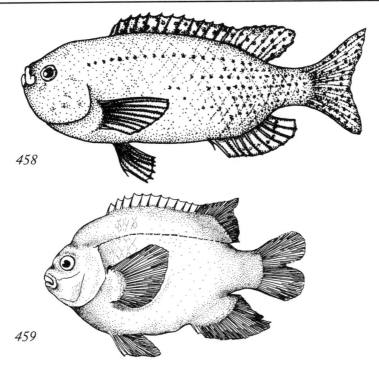

458

459

kelp beds. Specimens will reach 14 inches in length. The blacksmith is distributed from Monterey, California south to Magdalena Bay, Baja California.

Hypsypops rubicunda—the garibaldi (Fig. 459). This is considered to be the most beautiful of the local marine fishes. Adults are bright orange in color; the young may have spots of blue over the body. Specimens grow to about 14 inches in length and are known from Southern California south into Baja California. It can be seen locally along rocky shores especially swimming in or near kelp beds. Collection of the garibaldi is prohibited by law.

Family Mugilidae—the mullet family (Fig. 460). The lateral line faint or absent. The backbone is composed of 24-26 vertebrae.

Mugil cephalus—the striped mullet (Fig. 460). There are 2 dorsal fins which are widely separated; the anterior fin has 4 spines present. The upper surface is olive-gray in color

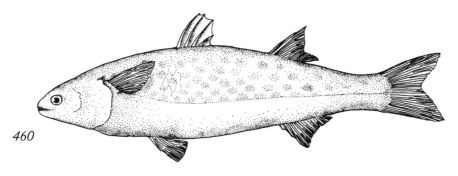

460

159

becoming lighter on the sides and undersurface. Specimens reach 3 feet in length and 15 pounds in weight. It is known from Central California to Chile especially in bays. It also occurs in Salton Sea.

Family Labridae—the wrasse family (Figs. 461, 462). These fish are characterized by their well-developed incisor teeth which protrude from the mouth. They are carnivores and cannabalistic in nature. Many of the species in the wrasse family bury themselves in the sand and sleep during the night.

Oxyjulis californica—**the senorita** (Fig. 461). The dorsal fin extends from the region of the gills to nearly the anal fin with the rays of nearly equal length throughout. Anal fin also extends for considerable length posterior to the anus. The lateral line dips sharply near the posterior limits of the dorsal fin (Fig. 461). This fish grows to about 10 inches in length and is brown in color. The senorita is found from Central California south into Baja

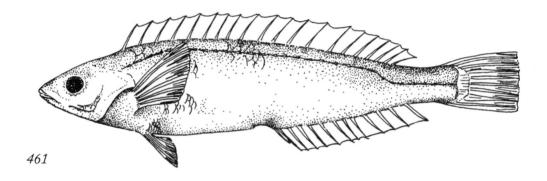

461

California and is present locally in inshore waters especially around kelp. It is caught by fishermen but is considered a nuisance.

Pimelometopon pulchrum—**the sheephead** (Fig. 462). The common name comes from the resemblance of its head to that of a sheep. It is distinctive in its appearance permiting easy identification. The head and posterior half of the male (Fig. 462) are darkly pigmented in contrast to a light colored to crimson mid-region. The lower jaw is white. The female is red to crimson in color. The sheephead will grow to 3 feet in length weigh up to 30 pounds. It is present from Central California south into Baja California. Locally it occurs around kelp beds off rocky shores. It is used by commercial fishermen as bait for lobster traps.

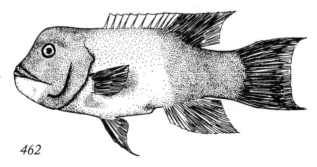

462

Family Blenniidae—the combtooth blennies (Fig. 463). The common name for this family comes from having jaws with comb-like teeth. The dorsal fin has more soft rays than spines.

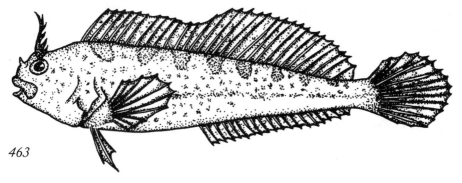

463

Hypsoblennius gentilis—bay blenny (Fig. 463). The flap above the eye is serrated along its posterior margin and not divided. The body is brown to green with reddish spots. It is a small species reaching 5-6 inches in length. It is found in bays in intertidal to 80 feet in depth from Monterey, California, into the Gulf of California.

Family Clinidae—the klipfish or scaled blennies (Figs. 464-466). The klipfishes are characterized by the possession of a long dorsal fin which extends to near the posterior end. The pelvic fins are small and located on the undersurface of the fish. Most of the species have a patch of short outgrowths over the nostrils and eyes.

Gibbonsia elegans—the spotted klipfish (Fig. 464). The spotted klipfish will grow to about 5 inches in length. The colors of *Gibbonsia elegans* are extremely variable; one phase is distinguished by 7 dark bars running from the upper to the lower surfaces; another phase has these bars as silver in color; a third color phase in plain. In all forms the ground color may vary from green to brown to orange. This species is known from bays, tide pools, to depths of about 100 feet offshore. It is known from California to Baja California.

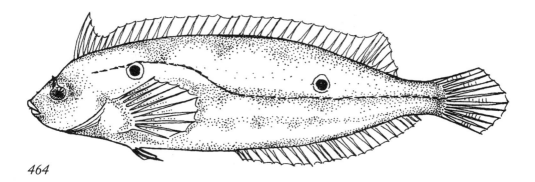

464

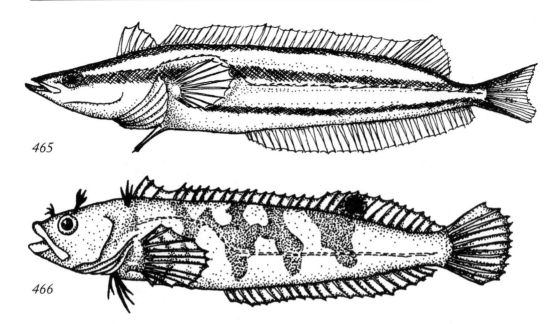

465

466

Heterostichus rostratus—the kelpfish (Fig. 465). The color of this fish varies and it can modify its mottled color to fit its background. The ground color varies from orange to brown and it may be mottled with a variety of lighter colors. The fish is known to reach 2 feet in length. The dorsal fin extends throughout most of the length of the fish and contains many spines. The anal fin extends through the posterior half of the body. The kelp fish is known from British Columbia to the southern tip of Baja California. It is especially common in kelp beds along inshore waters. Specimens enter bays and may be caught by fishermen.

Paraclinus integripinnis—reef finspot (Fig. 466). Spines are present the entire length of the dorsal fin. A large black spot (ocellus) present in the posterior third of the dorsal fin; this spot is the origin of the common name of this fish. The body is olive green with small dark spots. The length of the fish is 6 inches. It is present in intertidal to subtidal waters from Southern California to Baja California.

Family Stichaeidae-the prickleback fishes (Fig. 467). These are eel-like fish which lack pelvic fins and soft rays to the dorsal fin. There are 2 lateral lines on either side of the body.

Xiphister mucosus—the rock-eel (Fig. 467). The dorsal, anal and tail fins are continuous, with one another. The fish is green to black in color with a lighter cast on the lower surface. The rock-eel will grow to lengths of 20 inches. It is known from Alaska to Southern California especially in the intertidal zone.

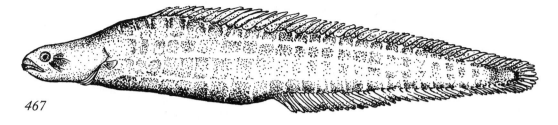

467

Family Gobiidae—the goby family (Figs. 468-471). The goby family is characterized by having their pelvic fins completely joined to each other. These joined fins form a type of sucker. Most of the species of gobies are small in size and live in the burrows made by shrimp, clams and other animals.

Clevelandia ios—the arrow goby (Fig. 468). This small species measures 2 inches in length. It is mottled gray-black in color on the upper surface and lighter in color on the lower surface. This fish is known to live in burrows of the echiuroid worm *Urechis caupo* (Fig. 86), a species rare in Southern California waters. The arrow goby undoubtedly inhabits the burrows of the ghost shrimp (Figs. 322-324). The arrow goby has been taken from the subtidal bottom with dredges from Los Angeles-Long Beach Harbors, Alamitos Bay and Newport Bay. *Clevendia ios* is distributed from British Columbia to Southern California.

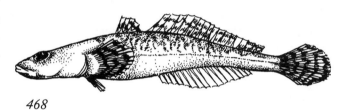

468

Gillichthys mirabilis—**the mudsucker** (Fig. 469). The mudsucker has a very large mouth with the upper jaw extending back as far as the gills. The color of this fish is mottled with darker shades over a brown-olive ground color. The lower surface is lighter in color. It will grow to 4 inches in length. *Gillichthys mirabilis* is known from California and is taken especially in Alamitos Bay, Anaheim Bay and Newport Bay. It is an important fish for sport fishing in Southern California.

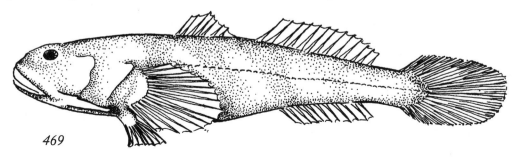

469

Ilypus gilberti—cheekspot goby (Fig. 470). The color of the cheekspot goby is tan to light gray. The common name is derived from the blue-black spot on the operculum. This small

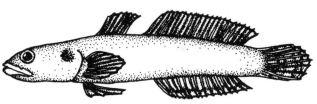

species measures up to 2.5 inches in length and is found in mud flats of bays from Northern California into the Gulf of California. This species was named in honor of a former principal of Los Angeles High School who worked with the David Starr Jordan in his studies of fishes of California.

470

Lythroypnus dalli—blue banded goby (Fig. 471). One of the more colorful species of fish in Southern California. This small species (2.5 inches in length) is red with blue bands. It

is found at rocky shores from the intertidal zone to over 200 feet. It is especially common at the Channel Islands. The bluebanded goby is found from Morro Bay, California into the Gulf of California.

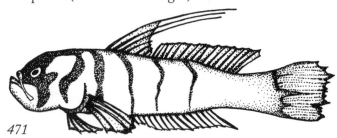

471

Family Cybiidae—the Spanish mackerel family (Fig. 353). There are 2 dorsal fins with the anterior one long with the spines becoming shorter posteriorly. Body completely covered with scales including the head.

Sarda lineolata—the **Pacific bonito** (Fig. 472). Small finlets are present posterior to the dorsal and anal fins. The upper surface is blue with a metalic luster with the lower surface becoming lighter in color. Oblique darker stripes extend from the lateral line. The Cali-

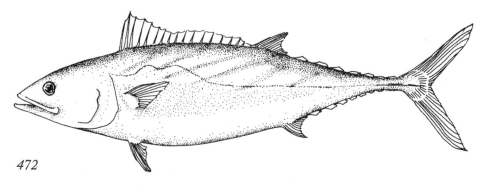

472

164

fornia bonito grows to lengths of 40 inches and is found from British Columbia south into Baja California. It is found in offshore waters and will extend into protected waters such as Lower San Gabriel River.

Family Bothidae—the lefteyed flounder fish (Figs. 473-475). The common name for this group of flatfish comes from the fact that the right eye migrated to the left side of the fish during its early growth and development. Right or left eyed flatfish can be distinguished from one another by holding the fish in such a way that the mouth is away from you and the dorsal fin upward; the side which the eyes are located indicates whether it is right or left eyed.

Citharichthys sordidus—the **Pacific sanddab** (Fig. 473). The lateral line is straight extending from the operculum to the tail fin. The scales are large and loosely attached. The Pacific sanddab measures up to 16 inches and is brown in color with occasional orange or black spots. It is widely distributed from Alaska to Baja California. It is present in subtidal offshore waters. It is of minor importance commercially.

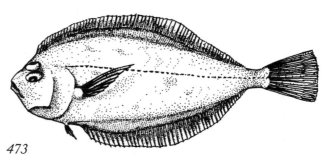
473

Citharichthys stigmaeus—speckled sanddab (Fig. 474). The common name of this species of flatfish is the most characteristic feature of the fish; it is speckled with black over the

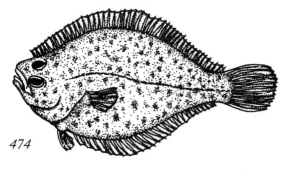
474

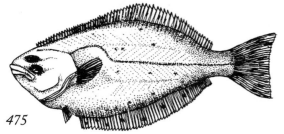
475

upper surface. It has been collected from Alaska south to Magdalena Bay, Baja California from subtidal depths to over 1000 feet.

Paralichthys californicus—the **California halibut** (Fig. 475). This fish may reach 5 feet in length and weigh 60 pounds. The lateral line arches over the pelvic fin. The upper surface is dark and frequently mottled and the lower surface white. This species ranges from Central California south and into the Gulf of California; it is found in offshore waters but smaller specimens occasionally enter bays. The California halibut is of importance both commercially and as a sport fish.

Family Cynoglossidae—the tonguefish (Fig. 476). A flatfish in which the dorsal and caudal fins are continuous with the anal fin. The anal fin is pointed.

Symphurus altricauda—**California tongue fish** (Fig. 476). The common name is derived from its shape which resembles a tongue. This species lacks a lateral line. It is light brown to gray with mottled spots on the upper surface and white on the lower surface. It is common on the bottom in subtidal, offshore waters of Southern California. It reaches 8-9 inches in length. The California tonguefish is known from Northern California to Cabo San Lucas, Baja California.

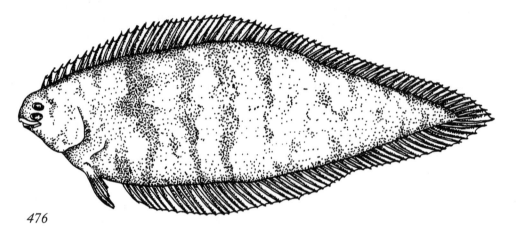

476

Family Pleuronectidae—the right eyed flounder family (Fig. 477-478). This family is so named because the left eye migrated to the right side of the body during its early growth in contrast to the left eyed flounder family discussed above.

Hypsopsetta guttulata—**diamond turbot** (Fig. 477). The body outline of this flatfish is diamond shaped. It is light gray in color with light blue spots on the upper surface. It lives on the bottom of bays and offshore waters where it feeds on benthic invertebrates. Specimens reach 18 inches. It is known from Northern California to Magdalena Bay, Baja California.

Micrustomus pacificus—**Dover sole** (Fig. 478). This common species in Southern California has a brown body with darkened fins. The mouth is extremely small in this species of flatfish which can reach 30 inches in length. It is present in bays and offshore waters from shallow depths to 3000 feet. It has been collected from the Bering Sea to Baja California.

Parophrys vetulus—**the English sole** (Fig. 479). The lateral line is nearly straight with a slight arch over the pelvic fin. The head is projected forward; lower jaw well developed. Upper surface brown and lower surface yellow or white. The English sole is distributed

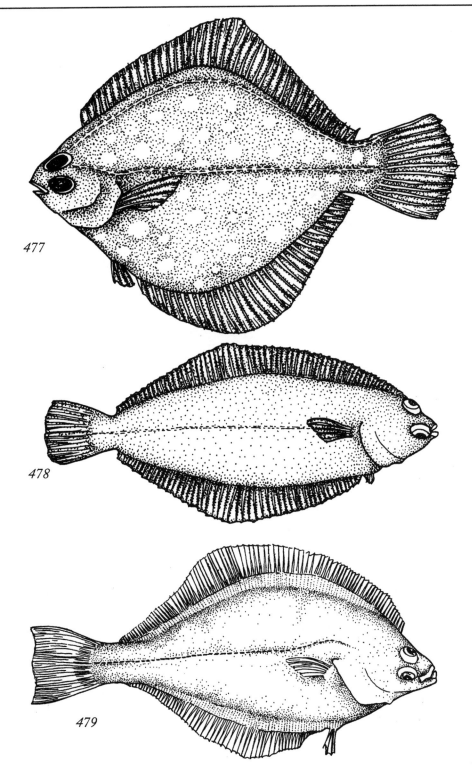

477

478

479

from Alaska to Baja California. It grows to about 20 inches in length. This species is found living on the bottom in offshore waters. It is an important commercial fish.

Family Cichlidae—the cichlids (Fig. 480). The lateral line is not continuous; it is interrupted back of the pectoral fin.

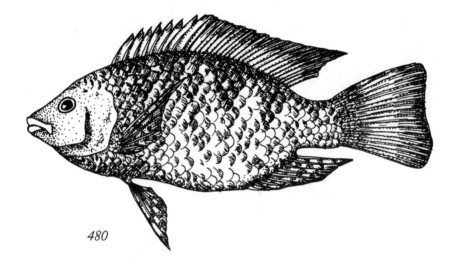

480

Tilapia mossambicus—tilapia (Fig. 480). This species does not occur naturally in Southern California. This species of *Tilapia* was originally from Africa. The different species of the genus *Tilapia* are cultured in many places of the World including United States. *Tilapia mossambicus* is a mouthbreeder; eggs are laid in a nest where fertilization takes place. The developing embryos are then carried in the mouth of the female (in some species the male may provide this function) until the young begin feeding. It was introduced into the fresh water reaches of San Gabriel River. It spread to the lower San Gabriel River where it is extremely abundant. It will probably spread to other estuarine rivers of Southern California. The correct generic name is still in question; some workers place it in the genus *Orechromis* rather than *Tilapia*. It is cultured as food in many parts of the world.

CLASS AVES—the birds (Figs. 481-514). The birds are warm blooded animals capable of flight and with feathers covering the body. They are highly variable animals with many parts of their bodies, such as their bill, wings, body shape and legs, modified for their environment. Many species of shorebirds and their eggs are protected by law. The food habits of the shore birds of Southern California are poorly known because of protection and because of the reluctance to kill a bird to determine its food habits. The eating habits of the shore birds is based on direct observation during feeding, analysis of fecal material, dead birds and from analysis of stomach contents of the same species from other geographical areas.

Family Podicipedidae—the grebes (Fig. 481). Aquatic birds which resemble ducks but differ in a lack of distinct tail feathers. They skim the water while taking flight. They dive or submerge during feeding.

Aechmorphorus occidentalis—the **Western grebe** (Fig. 481). Males and females are similar in appearance. The upper surface is black and the lower surface white. The bill is light yellow and straight. It measures 27-29 inches in height. It is found from Canada to Mexico. It breeds in Canada to Southern California laying its eggs in a nest in reeds. It winters along the Pacific coast and inland. *Aechmorphorus occidentalis* feeds on a variety of swimming invertebrates, fish and tadpoles. Its voice is a shrill whistle. Locally the Western grebe is found in Anaheim Bay, Newport Bay and offshore waters.

Family Phalorrocoracidae—the cormorants (Fig. 482). Cormorants have a neck which is in the shape of an "S". The bill is slender with a hook at the end.

Phalocrocorax auritus—**double crested cormorant** (Fig. 482). This species is characterized by a yellow-orange colored throat. The crest is not very evident in this species. It is widespread in United States and overwinters in Central America.

Family Alcidae—the auks (Fig. 483). This family has short necks and short stubby bills. They are excellent swimmers.

Ptychoramphus aleuticus—**Cassen's auklet** (Fig. 483). The feathers are dark except for white in the stomach region. The lower jaw has a white spot. Adults measures about 8 inches in height. It is distributed in the coastal regions of Alaska south to Baja California. It is very common bird especially in the Channel Islands.

Family Anatidae—the swans, geese and ducks (Fig. 484). A large group of

481

482

483

484

water birds characterized by webbed front toes and serrated margins along the cutting edge of the bill. These birds are swimmers and primarily migratory.

Melanetta perspiculata—**the surf scooter** (Fig. 484). The male is black with 1-2 white blotches of white on the crown or nape and the female is brown with 2 white spots on the side of the head. Adults measure about 20 inches in length. The adults rarely make a sound. The surf scooter is found from Alaska to Baja California. In Southern California it can be seen along any of the sandy beach areas as well as Anaheim Bay and Newport Bay. It makes its nest in bushes and lines it with down.

Family Procellariidae—shearwaters (Fig. 485). The shearwaters resemble gulls and live at sea. The nostrils are tube-like and located on the top of the bill.

Puffinus griseus—**sooty shearwater** (Fig. 485). It is dark colored with whitish linings on the undersurface of the wings. It is 17 inches tall. It breeds in Australia, New Zealand and South America and is found in North America during the summer. The sooty shearwater and other members of the family feed on fish, squid and crustaceans.

485

Family Pelecanidae—the pelicans (Fig. 486). Characterized by the long, flat bills with long throat pouches which can be inflated when full of food.

Pelecanus occidentalis—the brown pelican (Fig. 486). Males and females are similarly marked. They are brown in color except with a whitish head and neck. The wing span reaches over 50 inches in length. It is capable of plunging into the water from flight and capture fish or crustaceans. The adult does not make any noise. The brown pelican is distributed from the coastal areas of United States south into South America. In Southern California it is found in outer Los Angeles-Long Beach Harbors, Anaheim Bay, the mouths of rivers, Newport Bay and other localities. It breeds on the offshore islands where it nests on the ground.

486

Family Laridae—the gulls and terns (Figs. 487-493). This family consists four subfamilies, two of which are included: the Laridinae and the Sterninae.

Subfamily Laridinae—the gulls (Figs. 487-491). The species in this subfamily are long-winged swimmers with small feet which are webbed. Gulls are distinguished from terns by hooked bills (Fig. 487) compared to the straight ones in terns (Fig. 492). Gulls take from two to four years to obain their adult plumage. The western gull is an example of a species that requires four years, Heermann's gull three years and Bonaparte's gull needs only two years to reach its adult plumage.

Larus occidentalis—the western gull (Fig. 487). Adult plumage colors acquired during the third year. The wings and back are dark in color with a white head and breast. The bill is yellow with a red spot on the lower mandible. Juvenile plumage is mottled gray. Adults are about 2 feet in height. The western gull, as other gulls, feeds upon any available food, plant or animal. It makes a guttoral characteristic sound. The western gull is distributed from British Columbia south to Baja California. This species is widely distributed along the Southern California coast including rocky shores, sandy beaches and bays and harbors. It nests on sea cliffs or offshore islands.

487

488   489

Larus heermanni—**Heermann's gull** (Fig. 488). The adult has a white head with a red bill and a black tail. It stands about 20 inches. It is distributed all along the Pacific coast of North America but breeds on the offshore islands of Baja California.

Larus californicus—**California gull** (Fig. 489). The lower jaw has a red and black spot present. The legs are yellow-green in color. It will reach 23 inches in height. The California gull is widely distributed along the seacoast, lakes and farm lands of western North America.

Larus delawarensis—**ring-billed gull** (Fig. 490). Resembles the California gull but is smaller standing only 19 inches in height. The ring-gilled gull is distinguished from the California gull by the presence of a black ring around bill. Its hatitat is also similar to the California gull. It is found in western Canada and United States and winters in Mexico.

Larus philadelphia—**Bonaparte's gull** (Fig. 491). A small gull in which adults reach 13 inches in height. The head is white with a dark spot behind the ear and black during the summer. It is found in Alaska and western Canada during the summer and western

490

491

United States and Mexico during the winter. It inhabitats marine, fresh water and wet-lands.

Subfamily Sterninae (Figs. 492, 493).   They are separated from the gulls by possession of a straight bill rather than one with a hook at the end and usually with forked tail. They do not swim; gulls do.

Sterna forsteri—**Forester's tern** (Fig. 492).   Easily recognized by its forked tail in flight. It has gray colored back with white on the lower parts of the head and neck and breast. During non-breeding season (Fig. 493) there are black feathers around the eyes and the bill is black; during breeding season the head becomes black and the proximal part of the bill orange-red in color. It stands 1.5 feet in height. This species makes a variety of sounds includ-ing a nasal "zap". Forester's tern is widely distributed from Canada and Washington to the Atlantic coast south to Guatemala. It mi-grates through the mid-West and winters in California where it is found on both fresh water lakes and salt water bays. In Los Ange-les-Orange Counties *Sterna fosteri* is found near the coastline and within such bays as Anaheim and Newport. It builds its nest in marsh vegetation.

492

Sterna antillarum—**least tern** (Fig. 493).   Its bill is yellow in the summer and black during the winter months. The least tern has yellow legs and white forehead. It reaches 9 inches in height. It nests on bare gound in different areas of Southern California such as the Venice beach and upper Newport Bay. Islands in Anaheim Bay and Newport Bay have been constructed as mediating measures to provided protected nesting sites for this species. The least tern is an endangered species which shows signs of improving. It occurs in the coastal areas of United States and overwinters south of the border.

493

Family Ardeidae—the herons and bitterns (Figs. 494-495).   Members of this family are large and adapted for wading; they have long narrow legs and long spear-like bills. These birds feed on fish, frogs and aquatic invertebrates.

Ardea herodias—**the giant blue heron** (Fig. 494).   The sexes are similar in color. Specimens reach 4 feet in height. The feathers are blue-gray in color along the body and whitish on the head. It makes a deep croaking sound. It is widely distributed in North America where it is found in marshes, swamps or similar habitats. It can be seen locally in Anaheim Bay and upper Newport Bay. It builds a nest of sticks in reeds near marshes.

Leucophoyx thula—**the snowy egret** (Fig. 495).   The sexes are similar in appearance. A smaller sized member of the family, measuring slightly over 2 feet in length, its white color, with many emerging fine feathers from the body especially during breeding season and with black legs and yellow feet. It makes a low croaking sound. The snowy egret is widely found from

494

495

North America south to Argentina. The nest is made of sticks and is found among reeds or bushes. In Southern California the snowy egret is known from Anaheim Bay and upper Newport Bay.

Family Threskiornithidae—ibises and spoonbills (Fig. 496).   Many members of this family have bills that curve down, long legs and are waders.

Nycticorax nycticorax—**black-crowned night-heron** (Fig. 496).   This species is short and stocky and stands up to 28 inches tall. The top of its head is black and during breeding season it has two long head plumes. It is found in many parts of the world; in California it is found in the coastal regions. The black-crowned heron feeds on small fish, crustaceans, insects, worms and occasionally small birds and mammals.

Family Rallidae—the rails and coots (Figs. 497-499).   Marsh birds that are shaped like chickens.

Fulica americana—**American coot** (Fig. 497).   A duck-like bird that will stand 16 inches. Its head is black with a reddish patch in front. The bill is short,

496

175

497   498

stout and white in color. The feet are large and yellow. The American coot is very common in Southern California where it can be found in marine waters as well as lakes and ponds. It feeds on a variety of foods including aquatic plants and small animals.

Purzana carolina—sora (Fig. 498).   The sora has a yellow bill and a black face. It will grow to 9 inches in height. Sora feeds on crustaceans, insects and snails. It breeds throughout much of western Canada and United States and overwinters in the San Francisco Bay region, Southern California and south to Peru.

Rallus longirostris—clapper rail (Fig. 499).   This endangered species has a long bill that curves downward. It is tan in color with a white tail patch. The diet of the clapper rail is

499

500

varied including crabs, shrimp, small fish, insects and worms. It lives in estuarine marshes of San Francisco Bay and Southern California into Baja California.

Family Haematopodidae—the oystercatchers (Fig. 500). Their bill is long, flattened on the sides and red in color.

Haematopus bachmani—**black oyster catcher** (Fig. 500). This species appears black in color when standing (up to 17 inches), but white feathers are noted in flight along the back margins of the wings and tail. It feeds on snails, clams and worms. The black oyster catcher is a year round resident in the coastal regions of Alaska south to Baja California.

Family Recurvirostridae—the avocets and stilts (Fig. 501). The legs are long and slender; the bills are slender and curved upward in the avocets.

Recurvirostra americana—**American avocet** (Fig. 501). The American avocet stands up to 20 inches in height. The black and white color pattern and the slender, upturned

501

177

bill makes this species easy to identify. Its food consists of crustaceans, insects and other invertebrates. The American avocet breeds in Canada and winters in California to Central America. It is seen locally in upper Newport Bay and other wetlands of Southern California. It also inhabits lakes and ponds.

Family Charadriidae—the plovers, turnstones and surfbirds (Figs. 502, 503). This group of wading birds has short, thick necks and a compact body.

*Squatarola squatarola—***the black-bellied plover** (Fig. 502). The plumage is similar in both sexes but it varies from the breeding to non-breeding season. During the breeding season the breast is black and the back light colored. During the non-breeding season the breast is light colored and the back speckled gray (Fig. 502). They measure about one foot in height. Its voice is a slurred whistle. It breeds in the Arctic regions and overwinters from British Columbia to California. Locally it is found in Anaheim Bay and upper Newport Bay. It makes its nest in the tundra in the Arctic regions.

502

503

*Charadrius alexandrinis—***snowy plover** (Fig. 503). The male has a slim, black bill and a dark ear patch which may be lacking in the female. This small bird is 6 inches in height. It feeds on small invertebrates and vegetation. It is widely distributed in many parts of the world; in the west it is a year round resident on the beaches and sand flats of the Pacific Coast.

Family Scolopacidae—the snipes and sandpipers (Figs. 504-513). This group of wading birds has a slender body and a narrow neck. Their bills are long and slender.

Catoptrophorus semipalmatus—**the willet** (Fig. 504). The sexes are alike in plumage. At rest the breast is whitish in color and the back is gray, but during flight 2 white and black stripes can be seen on the wings which makes the willet easy to identify at this time. It measures about 1.5 feet in height. It makes a variety of musical sounds. The willet is found from Canada to Brazil. It winters along the coast of California in both fresh water and marine marshlands. It will make its nest in the grass. In Southern California it can be found in Anaheim Bay and Newport Bay.

Crocethia alba—**the sanderling** (Fig. 505). During breeding season the plumage of the sanderling is rusty gold along the back and lighter beneath. During non-breeding season it is pale gray above and light beneath. It measures about 8 inches in height. Its sound is short and distinctive. The sanderling migrates from the Arctic south into many areas of the Northern Hemisphere and into the Southern Hemisphere. It is found in groups along the sandy beaches where it moves up and down with the tides. It lays its eggs on the tundra.

Limosa fedosa—**the marbled godwit** (Fig. 506). This shorebird is easily recognized by its long, upturned bill. The plumage is a mottled buff-brown in both sexes. The marbled godwit measures up to 20 inches in height. It

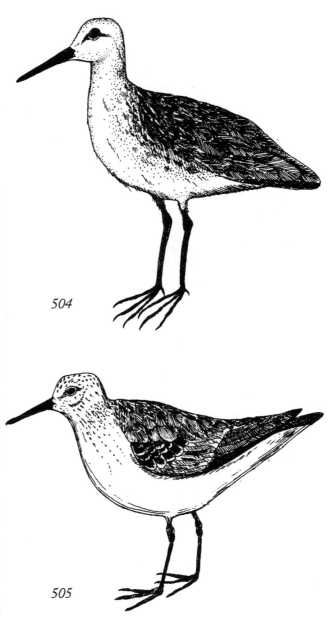

504

505

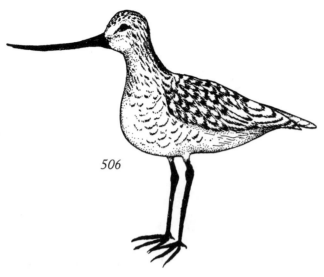

506

makes a sound like "godwit" which accounts for its common name. It migrates from its breeding area of the Great Plains area of United States and Canada down the Oregon and California coasts. Locally it is found in back bay areas of Anaheim Bay and Newport Bay and along the sandy beach shores. It lays its eggs in the grass.

Numenius americanus—the long-billed curlew (Fig. 507). Readily distinguished from the marble godwit in that its long bill is turned downwards. Its plumage is brown speckled along the back and a brown breast. It reaches a height of about 2 feet. The voice is a long "curlee" sound and other sounds. The long-billed curlew is distributed from Canada to Guatemala. It breeds in the north plains area and overwinters in the south. In Southern California it can be seen in tidal flats such as Anaheim Bay and Newport Bay. It builds its nest in a hollow on the praire.

507

508   509

Numenius phaeopus—**whimbel** (Fig. 508). Similar in appearance to the long-billed curlew but is only 1.5 feet tall. It plumage is grayer. Both species have the down curved, long, slender bill. The voice is different and is short "ti-ti-ti-ti". It feeds are a variety of invertibrates. It breeds in the Arctic and winters south to South America. Locally it is found on the mud flats of Bosa Chica and upper Newport Bay.

Tringa melanoleuca—**greater yellowlegs** (Fig. 509). As the common name suggests, it is characterized by long slender yellow legs. Its back is gray and checkered with brown, black and white. In flight the wings are dark with white tail feathers. It stands 14 inches tall. The bill is long and slightly upturned. The food habits are similar to the other members of the sandpiper family. Greater yellowlegs lives in both freshwater and marine marshes and mudflats. It is found in Alaska and Canada during the summer, and winters in the lower 48 states to South America.

Limnodromus scolopaceus—**long-billed dowitcher** (Fig. 510). The bill is long and straight. This 1 foot tall bird is grayish-brown in color with lighter plumage underneath. Breeding birds take on a rusty color. It migrates mostly along the coast from its summer locality in Alaska and Canada to the Washington to Central America in the summer. It lives in marshlands and mudflats favoring fresh water more than salt water.

510

511   512

Calidris alpina—dunlin (Fig. 511). Less than 1 foot tall with a long, stout, bill that is slightly curved downward. In the summer it is rusty red on its upper surface and a large black patch across the breast. Winter plumage is gray-brown above and grayish underneath. The dunlin is widely distributed in the Northern Hemisphere. In North America it breeds in the Alaskan-Canadian Arctic and in the winter it is found all along the Pacific Coast into Mexico. Locally it is found on the mudflats where it feeds on invertebrates.

Calidris minutilla—least sandpiper (Fig. 512). The least sandpiper is only about 6 inches in height. The upper surface is brown and the breast white. The bill is shorter than many of the species in this family. It breeds in Northern Alaska and Canada and spends the winter along the Pacific Coast to South America. It is often found in marshy areas as well as mud flats.

Actitis macularia—spotted sandpiper (Fig. 513). In the summer it is recognized by its dark spots on whitish plumage of the breast; these spots are not present during winter. Another characteristic feature is the white triangular wedge near the shoulder. The spotted sandpiper stands a little over 7 inches. It breeds throughout much of western United States, Canada and Alaska. In the winter months it is found in California to South America. It is found in both freshwater and at seashores.

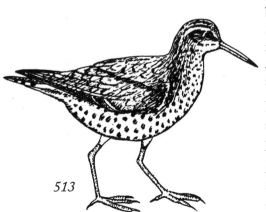

513

Family Fringillidae—sparrows, finches and grosbeaks (Fig. 514). Some authors place these species in Family Emberizidae.

Passerculus sandwichensis beldingi-Savannah sparrow (Fig. 514). There are many subspecies of *Passerculus sandwichensis* which are found throughout North America. The

514

subspecies *beldingi* is found from Santa Barbara County south to Bahia de San Quintin, Baja California. It is a small bird measuring less than 6 inches in height. Belding's Savannah sparrow is brown to beige in color with lighter tones underneath. There is a characteristic yellow tinge between the eye and the bill. It lives and nests in the upper tidal reaches of salt marshes where there is an abundant growth of pickleweed. They feed on insects when plentiful and on pickleweed during the winter months. This species was placed on the endangered species list of California in 1974.

CLASS MAMMALIA—the mammals (Figs. 515-521). These warm-blooded animals are characterized by having hair and nurse their young. The early mammals evolved on land and later invaded the ocean.

A total of 39 species of marine mammals has been reported from Southern California waters. Some of these have only been seen once or twice. Some of the mammals are permanent residents of the local marine waters; others migrate through the area on their way north and south. Some of them pass close to shore which enables one to see them from the shore. Charter boats take passengers out to sea to view the whales during the migratory seasons. A representative selection of marine mammals are discussed herein.

Order Cetacea—the baleen whales (Figs. 515-518). The forelegs are modified into flippers and they lack visible hind flippers. The tail is horizontal and extended laterally forming the fluke.

Suborder Mysticeti—baleen or whalebone whales (Fig. 515). They are characterized by having 2 blowholes and rows of flexible material termed baleens. Baleens filter plankton from the water; they are capable of taking in large quantities of water at one time in order

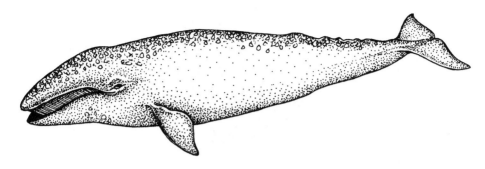

515

to filter out their food. There are several species of baleen whales which are known from Southern California, one of which is discussed which belongs to Family Eschrichtiidae.

Eschrichtius robustus—the California gray whale (Fig. 515). Male specimens will reach 50 feet in length. The head is small but the body is robust. It lacks a dorsal fin. The color of this whale is gray with light mottling and white streaks. It summers in the Bering and Chukchi Seas and winters in Baja California. It migrates southward through Southern California waters in December and January and northward in April through June. It feeds on crustaceans, mollusks and other invertebrates. It is an endangered species.

Suborder Odontoceti—the tooth cetaceans (Figs. 516, 517). This group has only one blowhold and teeth are present in one or both jaws.

Family Delphinidae—the dolphins (Fig. 516). These marine cetaceans have numerous conical teeth in both the upper and lower jaws. The tail has a notch located posteriorly.

Lagenorhynchus obliquidens—Pacific white-sided dolphin (Fig. 516). It is blue-gray with white streaks on the sides. It has a definite beak. This dolphin grows to 7 feet in length. The Pacific white-sided dolphin feeds upon schooling fish and squid. It ranges from Japan

516

to Baja California depending upon food availability. It is common in Southern California during the early fall months.

Tursiops truncatus—**Pacific bottle-nose dolphin** (Fig. 517). This species is the common one seen performing at sea aquaria. It is gray on the upper surface and a lighter gray beneath. It will reach 10 to 13 feet in length. It is found in temperate and tropical waters.

517

Family Phocoenidae—porpoises (Fig. 518). Porpoises lack a beak which distinguishes them from the dolphins. The dorsal fin is small.

Phocoenoides dalli—**Dall's porpoise** (Fig. 518). This species is very distinctive with its black and white body. The dorsal fin has a white tip and the sides of the body are white. Dall's porpoise grows to 7 feet in length. It feeds on crustaceans, squid and fish. It is known from Bering Sea to Baja California.

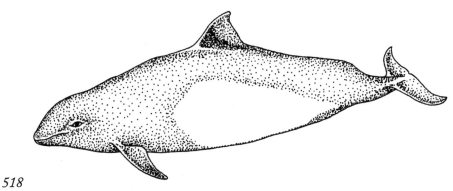

518

Order Carnivora—terrestrial and sea otters (Fig. 519). Primarily flesh eating mammals, such as the cat family, but others are omnivores and the giant panda is a herbivore.

Family Mustelidae—sea otters and some terrestrial species (Fig. 519). They have well-developed scent glands such as the skunk.

519

Enhydra lutris—sea otter (Fig. 519). The sea otter has small ears, hind feet webbed and flipper-like and the fur is soft and dense. It will reach 6 feet in length. At one time this species was distributed from Japan to Baja California, but heavy hunting nearly eliminated the species. While it is still on the threated species list, it has made a come back especially in Alaska. It is well known in Monterey Bay and occurs on some of the Channel Islands in Southern California. It has been filmed feeding on abalones and sea urchins.

Order Pinnipedia—includes the seals, sea lions and walruses (Figs. 520, 521). It has a short tail and the skin is thick.

Family Otariidae—the eared seals and sea lion (Fig. 520). These marine carnivores have small external ears and hind flippers which can be brought up along side its body.

Zalophus californianus—**the California sea lion** (Fig. 520). The male has dark brown to black fur, measures nearly 8 feet and weighs up to 1000 pounds. The female is lighter brown in color and is smaller in size. The California sea lion is found especially at rocky places off the Channel Islands of Southern California where it breeds. It migrates to British Columbia and southern Alaska during the winter. The California sea lion is often seen in circuses.

Family Phocidae—The earless seals (Fig. 517). These seals are separated from the Family Otariidae by the lack

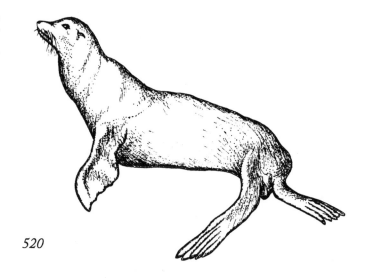

520

521

of external ears and the inability to bring their hind flippers forward. They are referred to as the "true seals".

Phoca vitulina—the harbor seal (Fig. 521). The fur of the harbor seal is coarse and has a characteristic spotted appearance. Specimens will reach about 6 feet in length. It feeds on fish, cephalopods and crustaceans. The harbor seal can be seen in offshore waters, and specimens will swim into protected waters. Over the years it has been seen in Marina del Rey, Los Angeles-Long Beach Harbors, Alamitos Bay and Newport Bay. It is widely distributed from China to Baja California and many localities in the northern Atlantic Ocean.

MARINE PLANTS

The majority of the species of marine plants encountered in the ocean belong to three seaweed phyla: Phylum Chlorophyta, or green seaweeds (Figs. 525-531), Phylum Phaeophyta, or brown seaweeds (Figs. 532-550), and Phylum Rhodophyta, or the red seaweeds (Figs. 551-571). These phyla usually can be distinguished from each other simply by color. All three phyla contain the green coloring matter found in most plants, termed chlorophyll, which is essential in carrying the processes of making plant food, or photosynthesis. The brown seaweeds have a brown pigment, known as fucoxanthin, which masks the green pigment chorophyll to give these seaweeds a brown color. The red seaweeds have a red pigment, known as the phycoerythrin which masks the green pigment chlorophyll to give these seaweeds a red color. The color of the brown and red algae may vary considerably depending upon the amount of the additional pigment present.

The flowering plants belong to Phylum Spermatophyta (Figs. 573-583). This group of plants, largely the grasses, have invaded the sea. Some of these can live entirely within the sea water while others live on the fringe, usually in marshes, between the ocean and land.

A very important group of plants are the diatoms. They are microscopic, but play a key role in the beginning of the food chain both in marine and fresh water. Diatoms are members of Phylum Chrysophyta. They live either in the water mass or attach to plants, animals or objects. Their cell wall is constructed of silican dioxide (sand) and their skeletons have formed extensive diatomaceous earth deposits. The cell is composed of two halves which fit together like a box. Diatoms may be centric in shape (Fig. 522), pennate (Fig. 523) or colonial (Fig. 524).

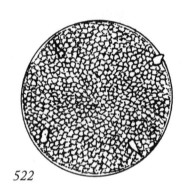

522

523

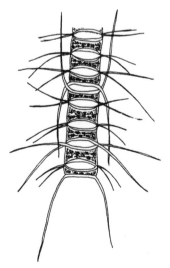

524

PHYLUM CHLOROPHYTA
The green seaweeds

These plants are grass green in color and lack any additional pigment to chlorophyll. They are frequently found high up in the tide horizon.

Family Ulvaceae (Figs. 525-527). The plants of this family are characterized by being only two cells in thickness. In the case of the genus *Ulva* the two cellular layers lie against one another and in *Enteromorpha* there is a hollow space making the species tubular. Many species of Family Ulvaceae occur in Southern California; only the larger, more commonly encountered species are included herein.

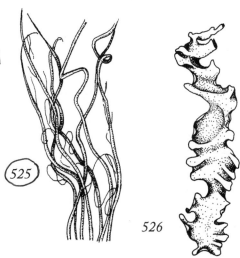

Enteromorpha crinita (Fig. 525). This species of green seaweed is tubular in shape and has many fine tubular branches. It forms extensive mat-like growths over the mud flat areas of Ballona Creek, Playa del Rey Lagoon, Alamitos Bay, Lower San Gabriel River, Anaheim Bay and Newport Bay. It is especially abundant in these areas in the spring and fall months. Other species of the genus *Enteromorpha* are present along some rocky shores and attached to boat floats in local marinas and harbors.

Ulva dactylifera (Fig. 526). This species of *Ulva* resembles sea lettuce except that the frond is long, narrow and frilled along the margins. It may grow up to one foot in length. Specimens have been collected along the jetty of Marina del Rey from the mid-tide horizon.

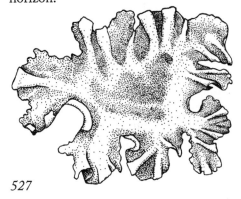

527

Ulva lobata—sea lettuce (Fig. 527). The common name of this seaweed comes from its resemblance to a lettuce leaf. This species, and several related species of *Ulva*, is used as food in other parts of the world. This species forms large, flat green sheets which may reach one foot in diameter. It is particularly abundant during the spring to fall months. Sea lettuce may be collected from boat floats in the protected waters at Marina del Rey, Los Angeles-Long Beach Harbors, Alamitos Bay, Huntington Harbor and Newport Bay as well as from mud flats present in back bays.

Family Cladophoraceae (Figs. 528-529). The plants are generally sessile and attached to the substrate with root-like structures. Members of this family are filamentous which may be branched (Fig. 528) or unbranched (Fig. 529).

✓*Chaetomorpha aerea* (Fig. 528). This green algae resembles <u>long grass</u>. The blades may at times reach one foot in length. It is bright green in color and can form extensive mats, along with *Cladophora trichotoma*, in high tide horizon tide pools. It has been collected from this niche at Lunada Bay, White's Pt., Pt. Fermin, Little Corona and Laguna Beach.

✓*Cladophora trichotoma* (Fig. 529). This is a small grass green alga which may reach 3 inches in length. It consists of many <u>branched tufts</u> and often forms extensive moss-like mats over the rocks in tide pools present in the high tide horizon. This species of *Cladophora*, and there are others in Southern California, has been collected at White's Pt., Pt. Fermin, the jetties of Alamitos Bay and Newport Bay, Little Corona and Laguna Beach.

530

Family Bryopsidaceae (Fig. 530). The plant is a branched tubular filament lacking cell walls giving the appearance of a single continuous structure.

Bryopsis corticulans (Fig. 530). This species is a delicately, pinnately-branched alga which can occasionally reach 2 inches in length. It may grow singulary or in tufts. *Bryopsis corticulans* grows in a variety of habitats; it has been taken from the rocks in the upper mid-tide horizons at rocky shores of the Palos Verdes area and from boat floats in the Marina del Rey, Los Angeles Harbor, Alamitos Bay, Huntington Harbor and Newport Bay.

Family Codiaceae (Fig. 531). The individual branches are large and circular in cross-section. This plant has a definite macroscopic shape. The

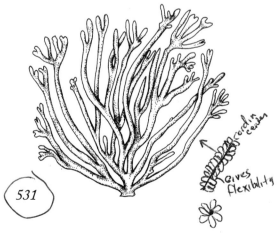

531

190

branches have a definite structure in cross-section when viewed under a compound microscope.

√ *Codium fragile* (Fig. 531).   This species is a dark, green, sponge-like alga which can form large branched clumps of a foot or more in length. The individual cylindrical branches measure about 0.5 inch in diameter. It occurs in the low intertidal level and is easily recognized. Clumps of *Codium fragile* have been seen at Lunada Bay, White's Pt., Pt. Fermin, Little Corona, Laguna Beach and on the jetty of Newport Bay.

PHYLUM PHAEOPHYTA
The brown seaweeds

The brown seaweeds are brown, olive-brown or dark green in color depending upon the amount of the brown pigment fucoxanthin present. This phylum includes the largest species of algae including the giant kelp which is an important local species.

Family Ectocarpaceae (Fig. 532).   Filamentous, branched plants in which the individual branch is only one cell wide.

Ectocarpus confervoides (Fig. 532).   This small brown colored seaweed is a delicately branched filamentous species which may reach 2-3 inches in length. Examination under the compound microscope shows the presence of numberous darker spherical to ellipsoid bodies which represent reproductive structures (Fig. 532). The individual cell walls may also be distinguished at this magnification. Many species in the genus *Ectocarpus* occur locally and specific identification requires a microscope and experience. *Ectocarpus confervoides* has been collected from boat floats in Marina del Rey, Los Angeles Harbor, Alamitos Bay and Newport Bay. Others species have been collected from rocky shores especially growing attached to other species of algae.

532

Family Dictyotaceae (Figs. 533-535).   Moderately large algae which are flattened and may or may not be branched. Growth is initiated by a single, microscopic cell at the tip of each growing edge or branch.

Dictyota flabellata (Fig. 533).   A brown colored alga which is more or less dichotomously branched to give it a fan-shaped growth form. The individual plant may extend several inches in length. The branches are thin and examination of a cross-section under the microscope will reveal a single layer of epidermal cells surrounding a single, interior layer of large cells. It is a common species along most rocky localities in Southern California where it is found attached to rocks in the low tide horizon.

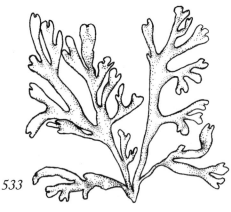

Dictyopterus zonarioides (Fig. 534). Similar in appearance to *Dictyota* it can be readily identified by the presence of a mid-rib in all its branches. Numerous fine hairs are present around its holdfast basal portions of the plant. It may grow to one foot in length and can be found in the low tide horizon attached to rocks at most rocky shores in Southern California.

Zonaria farlowii (Fig. 535). As the generic name applies, this species can be distinguished by concentric zones which radiate out from the holdfast. These lines can be readily seen with a hand lens. The branches are flat and brown in color and can reach nearly one foot in length. *Zonaria farlowii* is found especially in the mid-tidal level in sandy tide pools. Specifically, it has been collected at Lunada Bay, White's Pt., Pt. Fermin and Laguna Beach.

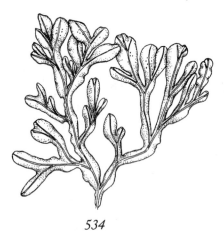

534

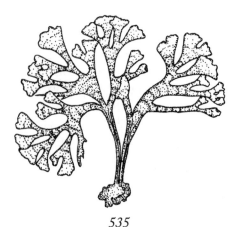

535

Family Desmartiaceae (Fig. 536). Moderate to large plants with pinnate branches from a single main shoot. The branches may be opposite or alternately arranged. The plant is flattened.

Desmarestia munda (Fig. 536). The yellow-brown plant is characterized by having opposite branches which are broadly elliptical in shape. The margins are off the main shoot and branches are serrated. Some specimens are known to reach 12-15 feet in length in colder

536

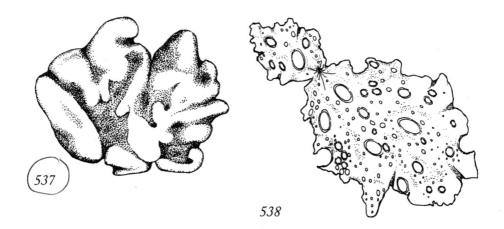

537

538

waters. This species has only been observed along the rock jetty of Anaheim Bay and Little Corona.

Family Punctariaceae (Figs. 537-539).   Small to moderate sized algae which often feels sponge-like to the touch. The growth form varies considerably in the different species in this family.

Colpomenia sinuosa (Fig. 537).   This small greenish-brown seaweed is subspherical in shape, hollow inside, and with many folds giving it a superficial appearance of a brain. It will grow to 1-2 inches in diameter. *Colpomenia sinuosa* grows in a variety of habitats. It can be found attached to rocks in the mid-tide horizon at any rocky shore locality; smaller forms grow attached to the feather boa *Egregia laevigata* (Figs. 542, 543) wherever it occurs, and attached to boat floats in Alamitos and Newport Bay.

golf ball size

Hydroclathrus clathratus (Fig. 538).   This peculiar light brown seaweed is characterized by being perforated with many small holes. It grows in a thin flat sheet, much like *Ulva lobata* (Fig. 527), which may be one foot in diameter. It does not occur too frequently locally, but some years it has been collected from the boat floats in Alamitos Bay and Newport Bay.

Scytosiphon lomentaria (Fig. 539).   This light brown species of algae is characterized by having many unbranched tubular structures which have many periodic constrictions. Under ideal conditions this species may grow to 1.5 feet

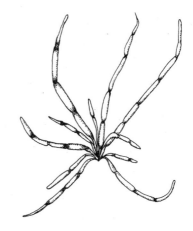

539

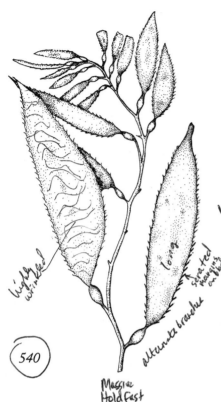

in length. It occurs throughout the intertidal zone but especially in the upper mid-tidal level attached to rocks. It has been seen at all rocky shores of the Palos Verdes Peninsula, Pt. Dume, Little Corona and Laguna Beach. It also occurs on the rock jetties at the entrance of Alamitos Bay and Newport Bay.

Family Lessoniaceae (Figs. 540, 542). The main stalk with numerous branches which may or may not be regularly branched. Plants can be small to large in size.

Macrocystis pyrifera—the giant kelp (Figs. 14, 540). This large olive-brown colored kelp is probably the most important species of algae in Southern California both ecologically and commercially. The plant occurs in offshore waters down to the depths of 60 feet although smaller plants occur intertidally. They grow to lengths of 100 feet. In certain localities it forms large underwater kelp forests (Fig. 14) which are important for many species of fish as well as other plants and animals. Kelp is harvested by boats which cut the plants off down about 4 feet from the surface of the water. The principle product is algin which is an emulsifying agent used in the preparation of commercial ice cream, chocolate milk, paints, cosmetics, pharmaceuticals, etc. The harvesting of kelp is regulated by the California Department of Fish and Game. The plant itself is recognized by the presence of a small air bladder at the base of each leaf-like structure. The most extensive beds of giant kelp in Southern California occur off Santa Barbara and Pt. Loma. The large holdfasts and fronds of this plant can be seen washed up on beaches especially after storms.

Pelagophycus porra—the elk kelp or bull kelp (Fig. 541). The one common name is derived from the two main branches coming off the air bladder (4-8 inches in diameter) each of which gives rise to smaller branches terminating in blades. The elk kelp is known to reach lengths of 90 feet. It is a subtidal brown which grows on the outer side of giant kelp beds. It is especially common on Catalina Island and the other Channel Islands rather than the mainland.

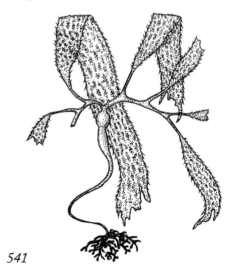

541

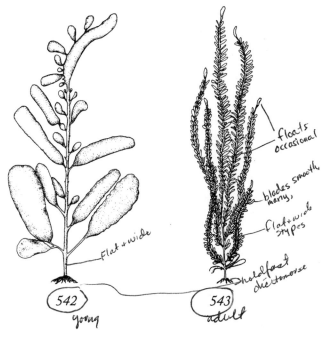

542 *young*

543 *adult*

(Handwritten labels in figure 542/543): floats occasional · blades smooth many · flat+wide stripes · Flat+wide · holdfast dichotomous · holdfast

Family Alariaceae (Figs. 542-544). The main stalk may or may not be branched. Lateral branches may occur all along the main stalk (Fig. 543) or at the tip (Fig. 544).

Egregia laevigata—the feathery boa (Figs 542, 543) This large brown colored seaweed attains lengths of 10-12 feet. The feathery boa attaches to rocks as a young plant and grows to about one foot in length as shown in Figure 542. This stage is generally found in the late fall through early spring months. With further growth the plant begins to take on its characteristic adult form (Fig. 543). The longer branchlets are each provided with an air bladder. *Egregia laevigata* is found attached to rocks at the low tide level at all rocky shores and most of the jetties in Southern California.

Eisenia arborea—the southern sea palm (Fig. 544). This brown colored alga is easily identified by its single stalk and a pair of branches containing many leafy blades. Intertidal specimens are generally 2 feet or less in height but luxurious subtidal specimens may grow several feet in length. Individual plants may live several years. The southern sea palm may be found in the low tide horizon at all rocky shores in Los Angeles-Orange Counties and also on the jetty at the entrance of Newport Bay.

Family Fucaceae (Figs. 545, 546). The branching is dichotomous and occurs in one plane.

(Handwritten labels in figure 544): one dichotomous branch · flat blades hang down · stiff · round short stype 1½" diam · very hard to break · sparser hold fast

544

Hesperophycus harveyanus (Fig. 545). This species can be distinguished from its closely related *Pelvetia fastigata* (Fig. 546) by the presence of a slight mid-rib and having faint spots on either side. The plant is more or less dichotomously branched and reaches about one foot in length. It occurs in the upper mid-tide horizon where it can form an extensive band at this tidal level, for example at Little Corona and Laguna Beach.

✓ *Pelvetia fastigata* (Fig. 546). [handwritten: *Rockweed*] The individual branches of this olive-brown plant are nearly cylindrical and may be swollen at the tip. As *Hesperophycus harveyanus* (Fig. 545), this species can also form extensive bands on the rocks, but they are generally lower in the mid-tide horizon. It is common at the rocky shores of Los Angeles-Orange Counties.

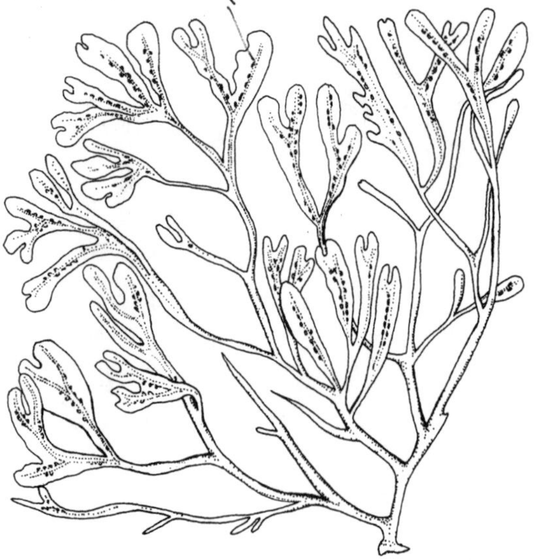

Family Sargassaceae (Figs. 547-550). Main branches primarily arise from the main stalk and are generally pinnately arranged. Terminal branches smaller and often have one (Fig. 547) or more (Fig. 548) airfilled vesicles present.

✓ *Cystoseira osmundaceae* (Fig. 547). This dark brown alga is characterized by having broad, fern-like leaves on the lower portion and delicate outer branches which have 4-10 small bead-like bladders in a line. This plant is known to grow to 25 feet in length in Central California but is much shorter in local waters. *Cystoseira osmundaceae* occurs at rocky shores from the low intertidal to subtidal depths. It has been observed along all the rocky shores of Palos Verdes Peninsula, the jetty at Newport Bay, Little Corona and Laguna Beach.

545

[handwritten annotations: ribs]

546

[handwritten annotations: Bumpy Lumpy + Hard; inflated; Dichotomus; Flatea; moundess; covers entire rock like shingles]

Halidrys dioica (Fig. 548). A closely related species to *Cystoseira osmundaceae*; it differs in that the bladders are flattened rather than round. The general form of both plants is similar in other respects. *Halidrys dioica* grows to several feet in length at rocky shores in the low intertidal and subtidal waters. It is not seen as often as *Cystoseira osmundaceae*; it occurs along all the rocky shores of Palos Verdes Peninsula and other rocky shores.

Handwritten: Sargasso grapes

Handwritten labels: Pinnate branching / bladder chains

547

Sargassum agardhianum (Fig. 549). The generic name comes from the Sargasso Sea which was named by the Portuguese sailors because of the presence of large floating masses of seaweeds of this genus. The original word was used for a type of grape; the small bladders present on the branches resemble grapes. In Southern California *Sargassum agardhianum* grows on rocks in the low tidal zone at all rocky shores.

Sargassum muticans (Fig. 550). This golden brown to dark brown seaweed grows up to 6 feet in length. It is separated from *S. agardhianum* by it alternate branches which are widely separated from one another. This species was introduced into the Puget Sound region about 1945 presumably by way of young oysters. It moved down coast and reached Southern California in 1970. It is especially abundant attached to rocks in the intertidal to subtidal rocks where the water is calm.

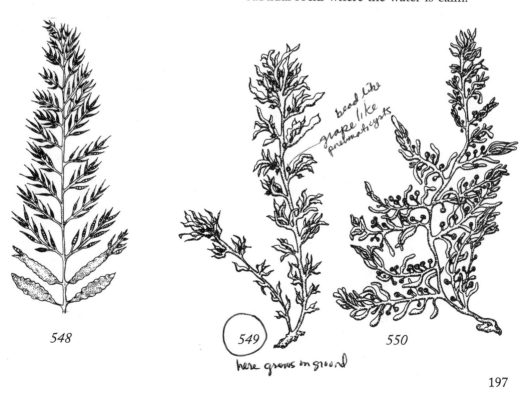

Handwritten labels: bead like / grape like / pneumatocysts

548

549

Handwritten: here grows on ground

550

197

Handwritten: In Atlantic + Floats in planktonic in gyres in huge amounts

PHYLUM RHODOPHYTA
The red seaweeds

The red color in these seaweeds is due to the presence of the red pigment phycoerythrin in addition to chlorophyll. Most of these seaweeds appear red in color but some species can be dark green in color.

Family Bangiaceae (Fig. 551).   Plant in a flat sheet-like structure which may be composed of a single or double layer of cells.

Porphyra perforata (Fig. 551).   This alga is one of those species which grows in broad, flat sheets. It can be readily distinguished from *Ulva lobata* (Fig. 527) and *Hydroclathrus clathratus* (Fig. 538) by its red color. The sheets may reach 1-2 feet in diameter. This species is one of several members of the genus which are cultivated in Japanese bays for food. *Porphyra perforata* has been collected locally from the jetties of Alamitos Bay and Newport Bay. Smaller species of the genus *Porphyra* may be collected at rocky shores in the upper mid-tide zone and attached to surf grass (Fig. 5).

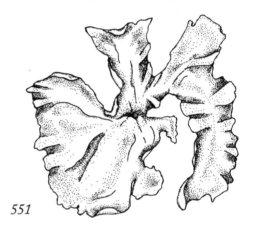

551

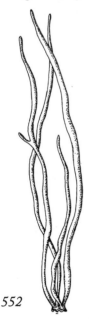

552

Family Helminthocladiaceae (Fig. 552).   Many cylindrical branches arising from a common base. The branches may either be gelatinous in texture or stiff as a result of secondary deposition of calcium carbonate.

Nemalion helminthoides (Fig. 552).   The specific name comes from its worm-like appearance. It differs from the other worm-like species, *Scytosiphon lomentarium* (Fig. 539), in having a sponge-like feel but lacking constrictions. Plants will grow to nearly one foot in length attached to rocks in the low tide horizon along the Palos Verdes Peninsula and Little Corona.

Family Gelidiaceae (Fig. 553).   Numerous branches arising in one plane from the main stalk. Other specific details of the family require the use of a compound microscope. Members of this family are important as the source for agar, a product used in the making of culture media for microbiology and in some foods and pharmaceuti-

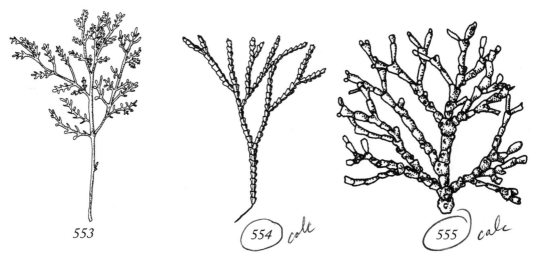

553
554 colc
555 calc

cals. Subtidal species of the genus *Gelidium* were harvested at one time in Southern California, but it is now collected primarily in Baja California with the refining done in Japan.

Gelidium coulteri (Fig. 553).   A small plant measuring 1-2 inches in height but a common one in mid-tide horizon. It has many fine branches and this species may form extensive mat-like beds. It is dark purple to olive-green in color and can be collected from all rocky shores and jetties in Southern California.

Family Corallinaceae—the coralline algae (Figs. 554-557).   Coralline algae are so named because their cells deposit calcium carbonate which gives them a hard covering. Coralline algae may either be jointed or articulated species (Figs. 553-557) or form solid mat-like growths (not shown, but some species occur on rocky shores).

Bossea **sp.** (Fig. 554).   Several species occur in Southern California and specific identification is difficult. This genus may be distinguished by its wing-like joints especially near the basal portions. They are pinkish-red in color and grow a few inches in height. It has been collected intertidally at all rocky shores in Los Angeles-Orange Counties and from the jetty at Newport Bay.

Calliarthron **sp.** (Fig. 555).   Species in the genus *Calliarthron* are heavier in appearance than the others. It occurs with *Bossea* but examination with a hand lens can usually distinguish the two genera; it lacks the wing-like joints characteristic of *Bossea*. It is pinkish-red in color and can form mats in the intertidal zone at all our local rocky shores.

Corallina **sp.** (Fig. 556).   There are several species of this genus locally with *Corallina vancouveriensis* being the most common. Members of this genus are distinguished by their pinnate branching (Fig. 556) especially

556   colc

at the ends. Members of this genus are pink and generally 1-2 inches in height. They can form extensive mats especially in the tide pools at rocky shores. Members of the genus *Corallina* can be seen at all rocky shores in Southern California, rock jetties, and on boat floats near the entrance of such areas as King's Harbor, Alamitos Bay and Newport Bay.

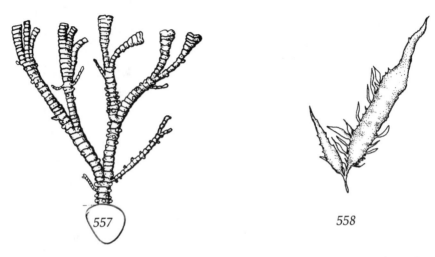

557

558

Lithothrix aspergillum (Fig. 557). This species is easily differentiated from the other jointed coralline algae by its unique appearing joints. It is also found along all local rocky shores in the intertidal zone.

Family Grateloupiaceae (Figs. 558, 559). Erect plants may be either as broadened sheets (Fig. 558) or branched (Fig. 559). Other characteristics require the use of a microscope to observe.

Grateloupia californica (Fig. 558). Plants in Southern California reach about 8 inches in length. This species is red to green-purple in color. Plants are found attached to rocks in the lower tidal zones at Little Corona and Laguna Beach.

Zanardinula lanceolata (Fig. 559). This bright red to purple alga is easily recognized by its tendency to grow in tufts below the California mussel bed zone especially where the wave action is strong. The branches may grow to about one foot in length. This species is known also under the name *Prionitis lanceolata*. It is present at all local rocky shores and also at the ocean end of the jetty at Newport Bay.

559

200

Family Nemastomaceae (Fig. 560). Plants erect and variously shaped. The central part of the branches is composed of many, microscopic, longitudinal filaments which requires a microscope to see.

Schizomenia pacifica (Fig. 560). The generic name is based on the tendancy of these species to become split (hence schizo meaning split) at the ends of the blades (Fig. 560). This species forms broad, flat plants which are red in color. Its characteristic growth form makes it easily identified. It is found in the low tide zone at all rocky shores but never common.

560

Family Solieriaceae (Fig. 561). This family may appear as a branched cylindrical plant (Fig. 561), or take a variety of shapes. Other characteristics are microscopic in nature.

Agardhiella coulteri (Fig. 561). This species may reach one foot in height; it is red in color. Main axis and branches appear cylindrical with knobby structures because of the presence of reproductive structures within.

Family Plocamiaceae (Fig. 562). Numerous branches especially at the tips where they are frequently curved. Branches generally in one plane.

Plocamium pacificum (Fig. 562). This red species is probably one of the more beautiful seaweeds of Southern California. Its delicate branching tends to be slightly curved at the

561

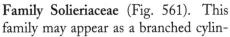
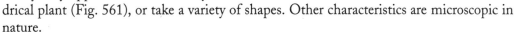

562

ends (Fig. 562). It grows fan-like and small pieces have been pressed on paper for use as place cards, notes, etc. It can be seen in tide pools at all local rocky shores down into subtidal depths. Pieces are often seen washed up on sandy beaches after storms.

Family Gigartinaceae (Figs. 563-567). This family contains many species in Southern California which take a variety of growth forms. Only a few of the more common species of Family Gigartinaceae are described herein.

Gigartina canaliculata (Fig. 563). This seaweed is olive to purple in color and will form dense mats of up to 3 inches in thickness over the rocks. It especially grows abundantly in the low tide horizon along all rocky shores. It may be found along the rocks of the jetties of Alamitos Bay and Newport Bay. This species may be distinguished from *Gelidium coulteri* (Fig. 563) by its fewer but larger branches.

Gigartina leptorhynchus (Fig. 564). A dark red to brown colored alga with numerous small branches arising all along the main stalks. Plants reach about 6-8 inches in height. It can be collected at low tide attached to rocks.

Gigartina spinosa (Fig. 565). This species of seaweed is characterized by having one or more broad flat blades each having few to several small branches arising from the margins. The surface of the blades are covered with numerous small fleshy outgrowths. *Gigartina spinosa* is greenish-purple to brown-red in color and may grow to 15 inches in length. It has been collected at low tide at the rocky shore localities of Palos Verdes Peninsula and Laguna Beach as well as the jetties of Alamitos Bay and Newport Bay.

Gigartina volans (Fig. 566). A large purple-brown plant which may reach lengths of 12-14 inches. Small branches arise laterally from the base giving way to a larger, flat blade which has

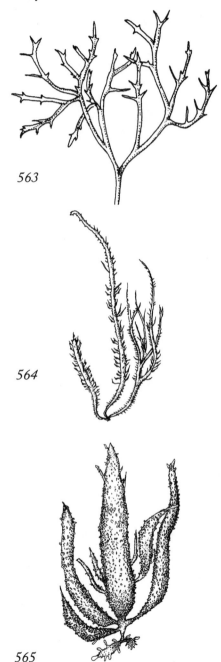

563

564

565

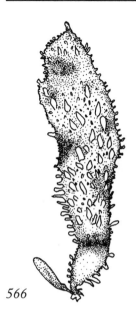

numerous small branches along its margin. It has been found during minus tides at Laguna Beach.

Rhodoglossum affine (Fig. 567). This species of seaweed is greenish to purple in color and grows up to 6 inches in length. The base is generally cylindrical in cross-section, but the branches are flattened. Branching is generally dichotomous. *Rhodoglossum affine* is found at low tides along the open coast wherever rocks are present.

Family Rhodymeniaceae (Fig. 568). Members of this family are highly variable in appearance and the family characteristics are based on microscopic details.

Rhodymenia pacifica (Fig. 568). This species of alga is pink to red in color. It is flattened and has dichotomous branching. The individual plant may reach about 5 inches in length. It can be collected from the low to subtidal depths at the jetty at Marina del Rey, Palos Verdes Peninsula, Little Corona and Laguna Beach.

566

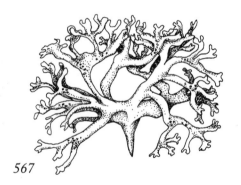

567

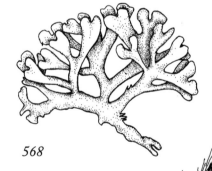

568

Family Ceramiaceae (Fig. 569). Fine, delicate red algae which have a filamentous structure.

Antithamion occidentale (Fig. 569). This finely branched red alga requires a microscope for positive identification. It could be confused with another red alga, *Polysiphonia pacificus* (Fig. 572), or other finely branched red species not described herein. This species occurs in protected waters attached to boat floats in Marina del Rey, Los Angeles-Long Beach Harbors, Alamitos Bay, Huntington Harbour and Newport Bay.

569

Family Rhodomelaceae (Figs. 570, 571).   Generally erect plants with many branches. The tips of which have many, microscopic branched filaments.

Laurencia pacifica (Fig. 570).   A reddish-purple algal species which is usually a few inches or less in length. Many branches arise from the main stalk which in turn may also branch. This species of red alga grows attached to rocks at low tide. It has been collected at White's Pt., Little Corona and Laguna Beach.

Polysiphonia pacifica (Fig. 571).   This reddish, delicately branched seaweed resembles *Antithamion occidentale* (Fig. 570); however, under the compound microscope you will note that the main branches are

570

571

composed of many parallel cells in *Polysiphonia pacifica* (Fig. 571). This species may form growths up to 5 inches in length. It has been collected from the rocks, pilings and boat docks in Alamitos Bay and Newport Bay.

PHYLUM SPERMATOPHYTA
The flowering plants

These plants have true roots, stems and leaves and produce seeds which are formed as a result of sexual reproduction. This phylum contains most of the conspicuous land plants of the world.

CLASS ANGIOSPERMAE—the true flowering plants (Figs. 572-582).   These plants all bear flowers although some of them are difficult to see. They produce a seed enclosed within a dry or succulent fruit. Most of the conspicuous plants of the earth, with the exception of the conifers (evergreen trees), belong to this group of plants. The angiosperms evolved on land. Later some of them secondarily invaded the sea either as marsh plants in estuaries (Figs. 574-582) or true sea plants (Figs. 572, 573).

Family Zosteraceae—the eel grasses (Figs. 572, 573). This family includes the marine grasses which can be entirely submerged. The plants have creeping stems which help hold sands in place.

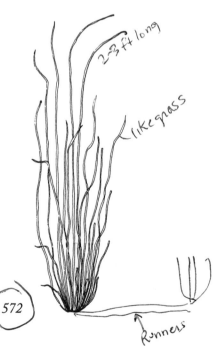

572

204

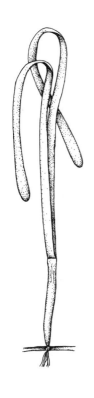

Phyllospadix scouleri-the surf grass (Fig. 572). The genus *Phyllospadix* is distinguished from *Zostera* (Fig. 573) by its narrow leaves and its habitat of rocky shores. A second species, *Phyllospadix torreyi*, occurs less frequently in Southern California. Separation of the 2 species of *Phyllospadix* is based on flower characteristics. *Phyllospadix scouleri* is abundant in the minus tide zone at all rocky shores of Southern California especially where the surf is strong. Its roots harbor many different species of worms, crustaceans, mollusks and echinoderms.

Zostera marina-eel grass (Fig. 573). This species of marine grass is recognized by its leaves which measure about 1/2 inch in width and its habitat of bays and estuaries. Local eel grass is not as important of a plant as it is in more northern bays; it is known to occur in small patches in Alamitos Bay, Anaheim Bay and Newport Bay where it is more prevalant. It grows subtidally at all localities where its roots extend into the sand or finer sediments.

Family Batidaceae (Fig. 574). Species of this family live near the ocean. Their leaves arise opposite one another and are fleshy. The flowers are formed at the end of a spike.

573

Batis maritina—the salt wort (Fig. 574). A low creeping plant in which the stems generally lie on the ground. The plant will grow on the ground and the main stems may extend out several feet. The salt wort is a coastal marsh plant which is known from Alamitos Bay, Anaheim Bay and upper Newport Bay.

574

Family Gramineae—the grass family (Figs. 575-577). This family contains the majority of the familiar grasses of the terrestrial environment. These plants generally have hollow stems and solid at the node (joints). The leaves are in two parts, the sheath which more or less remains attached to the stem and the blade wich extends outward from the stem.

✓ *Distichlis spicata*—**the salt grass** (Fig. 575). The most abundant salt marsh plant in Southern California. The plants form large patches, generally 3-4 inches tall, above the

high tide line and are rarely covered with sea water. The leaves are stiff and sharp to step on especially with bare feet. The salt grass can be found in estuaries which have largely been undisturbed by man such as parts of upper Alamitos Bay, Anaheim Bay, Bolsa Chica and parts of upper Newport Bay.

✓ *Monanthochloe littoralis*-the shore grass (Fig. 576). This grass is a creeping species which has short erect branches as shown in Fig. 576. The leaves arise in clusters and are sharp at the tip. The shore grass is found in most salt marsh areas of Southern California; it has been collected at Alamitos Bay, Anaheim Bay, Bolsa Chica and Upper Newport Bay.

✓ *Parapholis incurva*—**the sickle grass** (Fig. 577). An annual grass which grows to 5-6 inches in height. The leaves are long and narrow and curve around the stem for considerable length. The sickle grass has been found at Alamitos Bay, Anaheim Bay and Upper Newport Bay.

✓ *Spartina foliosa*—**the cord grass** (Fig. 578). This grass is the tallest of the various salt marsh plants; it can grow from 1-3 feet tall. Its leaves at the base curve around the stem and measure about 1/2 inch wide at this point. *Spartina foliosa* grows in extensive patches in the high tide horizon but low enough down so that it is frequently covered with a foot or more of sea water. It grows on the natural undisturbed mud flat areas of Anaheim Bay, Bolsa Chica and parts of Upper Newport Bay.

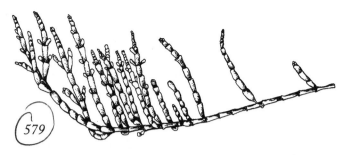

Family Chenopodiaceae—the goosefoot family (Fig. 579). Members of this family are especially adapted for growing in or near saline waters. The stems are frequently well developed and succulent with a resultant reduction in the size of the leaves.

Salicornia **sp.—the pickle weed** (Fig. 579). At least 2 common species of *Salicornia* are known from Southern California but they can only be separated by examination of the inconspicuous flowers. One species, *Salicornia virginica*, is perennial; the other species, *Salicornia bigelovii*, is an annual. The common name comes from its use as a pickled food or as a fresh vegetable in Europe. *Salicornia* is easily recognized by its succulent, jointed stems. It may be found in the undisturbed portions of upper Alamitos Bay, Anaheim Bay, Bolsa Chica and Upper Newport Bay.

Suaeda californica—**the sea-blite** (Fig. 580). A succulent plant found in the high tidal zone. It is a short-lived perennial which is able to invade bare sand dunes. The leaves are long and narrow and crowded at the stem. It grows either erect to spreading with the stems extending up to 3 feet in length. The flowers are greenish with the blooms appearing in July to October.

Family Plumbagomaceae—the thrift family (Fig. 581). Members of this family are able to secrete salt, and salt crystals are often present on the undersurface of the leaves.

Limonium californicum—**sea lavender or marsh rosemary** (Fig. 581). An erect perennial which will grow up to 3 feet in height which is found in high tidal zone. The reddish leaves are broad and flat and are often covered with salt. The flowers bloom in July to October and are pale lavender in color. The flowers will become dry and remain on the stalk.

Family Frankeniaceae (Fig. 582). Low growing perennial herbaceous plants with leaves arising opposite from

580

581

one another from the stem.

Frankenia grandifolia—**the alkali heath** (Fig. 582). The plant grows to 1 foot in height and is found in the high tidal zone. The oval shaped leaves are small and appear in bundles on opposite sides of the stem. The small pink flowers appear from June to October.

582

APPENDIX I

References to Pacific Coast Marine Life
General References

Allen, R. K., 1969. *Common Intertidal Invertebrates of Southern California.* Peek Publications.

Brandon, J. L. and F. J. Rokop, 1985. *Life Between the Tides.* American Southwest Publishing Co.

Dailey, M. D., D. J. Reish and J. W. Anderson, editors, 1994. *Ecology of the Southern California Bight.* University of California Press.

Emery, K. O., 1960. *The Sea Off Southern California.* John Wiley and Sons.

Gotshall, D. W. and L. L. Laurent, 1979. *Pacific Coast Subtidal Marine Invertebrates.* Sea Challengers.

Hinton, S., 1969. *Seashore Life of Southern California.* University of California Press.

Johnson, M. E. and H. J. Snook, 1927. *Seashore Animals of the Pacific Coast.* MacMillan. Reprinted in 1967 by Dover Publications.

MacGinitie, G. E. and N. MacGinitie, 1967. *Natural History of Marine Animals.* Second Edition. McGraw-Hill.

Morris, R. H., D. P. Abbott and E. C. Haderlie, 1980. *Intertidal Invertebrates of California.* Stanford University Press. [This reference includes color photographs of many of the common invertebrates of California. The text summarizes the known biological knowledge of the species. The chapters are arranged according to animal group.].

Rickets, E. R, J. Calvin and J. W. Hedgpeth, Revised by D. W. Phillips, 1985. *Between Pacific Tides.* 5th edition. Stanford University Press, Stanford.

Smith, R. I. and J. T. Carlton, 1975, editors. *Light's Manual: Intertidal Invertebrates of the Central California Coast.* 3rd edition. University of California Press, Berkeley.

References to Specific Groups of Organisms
These References Are Usually More Technical

PLANKTON

Dawson, J. K. and R. E. Piper, 1994. Zooplankton. In: Dailey, M. D., D. J. Reish and J. W. Anderson, editors. *Ecology of the Southern California Bight.* University of California Press, pp. 266-303.

Eppley, R. W., 1986. *Lecture Notes on Coastal and Estuarine Studies, Vol. 15. Plankton Dynamics of the Southern California Bight.* Springer-Verlag.

Hardy, J. T., 1994. Phytoplankton. In: Dailey, M. D., D. J. Reish and J. W. Anderson, *Ecology of the Southern California Bight.* University of California Press, pp. 233-265.

Newell, G. E. and R. C. Newell, 1963. *Marine Plankton. A Practical Guide.* Hutchinson Educational Ltd., London.

PHYLUM PORIFERA—The Sponges

Green, K. and G. Bakus, 1994. *Taxonomic Atlas of the Benthic Fauna of the Santa Maria Basin and the Western Santa Barbara Channel. Vol. 2, The Porifera.* Santa Barbara Museum of Natural History, Santa Barbara, CA.

Fry, W. G., editor, 1970. *The Biology of the Porifera.* Academic Press, London.

Laubenfels, M. W. de, 1932. The marine and Fresh Water Sponges of California. *Proceedings of the U. S. National Museum.* Vol. 82.

PHYLUM CNIDARIA—The Jellyfish, Sea Anemones, etc.

Fraser, C. McL., 1937. *Hydroids of the Pacific Coast of Canada and the United States.* University of Toronto Press.

Hand, C., 1954-55. The Sea Anemones of Central California. *Wasmann Journal of Biology.* Vols. 12, 13.

Mackie, G. O., editor, 1976. *Coelenterate Ecology and Behavior.* Plenum, New York.

PHYLUM PLATYHELMINTHES—The Flatworms

Hyman, L. H., 1953. The polyclad flatworms of the Pacific Coast of North America. *Bulletin of the American Museum of Natural History.* Vol. 100.

PHYLUM NEMERTEA—The Ribbon Worms

Coe, W. R., 1940. Revision of the Nemertean Forms of the Pacific Coast of North, Central and Northern South America. *Allan Hancock Pacific Expeditions, University of Southern California.* Vol. 2.

PHYLUM ENTOPROCTA
(See Phylum Ectoprocta)

PHYLUM ANNELIDA

Class Polychaeta—Segmented Sea Worms

Blake, J., C. Erséus and B. Hilbig, 1994. *Taxonomic Atlas of the Benthic Fauna of the Santa Maria Basin and the Western Santa Barbara Channel. Vol 2, Oligochaeta and Polychaeta,* Part 1. Santa Barbara Museum of Natural History, Santa Barbara, CA.

Fauchald, K., 1977. The Polychaete Worms: Definitions and Keys to the Orders, Families and Genera. *Los Angeles County Museum of Natural History.* Science Series. 28.

Fauchald, K. and P. A. Jumars, 1979. A Study of Polychaete Feeding Guilds. *Oceanography and Marine Biology Annual Reviews.* Vol. 17.

Hartman, O., 1968. *Atlas of the Errantiate Polychaetous Annelids from California.* Allan Hancock Foundation, University of Southern California.

Hartman, O., 1969. *Atlas of the Sedentariate Polychaetous Annelids from California.* Allan Hancock Foundation, University of Southern California.

PHYLUM ECHIURA

Fisher, W. K., 1946. The Echiuroid Worms of the North Pacific Ocean. *Proceedings of the U. S. National Museum.* Vol. 96.

PHYLUM SIPUNCULA

Cutler, E. B., 1994. The Sipuncula, Their Systematics, Biology, and Evolution. Cornell University Press, Ithaca, NY.

Fisher, W. K., 1952. The Sipunculid Worms from California and Baja California. *Proceedings of the U. S. National Museum.* Vol. 102.

PHYLUM MOLLUSCA—The Chitons, Snails, Clams, etc.

Behrens, D. W., 1980. *Pacific Coast Nudibranchs.* Sea Challengers.

Bernard, F. R., 1983. *Catalogue of the Living Bivalvia of the Eastern Pacific Ocean: Bering Strait to Cape Horn.* Canadian Department of Fisheries and Oceans.

Keen, A. M. and E. Coan, 1974. *Marine Mollusca Genera of Western North America: A Illustrated Key.* Stanford University Press, Stanford.

MacFarland. F. M., 1966. Studies of Opistobranchiata Mollusks of the Pacific Coast of North America. *Memoirs of the California Academy of Sciences.* Vol. 6.

McLean, J. H., 1978. Marine Shells of Southern California. Revised Edition. *Los Angeles County Museum, Science Series.* No. 24.

Morris, P. A., 1960. *A Field Guide to Shells of the Pacific Coast and Hawaii.* Houghton, Mifflin.

Putman, B. F., 1982. Littoral and Sublittoral Polyplacophora of Diablo Cove and Vicinity, San Luis Obispo, County. California. *Veliger.* Vol. 24.

PHYLUM ARTHROPODA
Class Crustacea

Barnard, J. L., 1969. Gammaridean Amphipods of the Rocky Intertidal of California: Monterey Bay to La Jolla. *Bulletin of the U. S. National Museum.* No. 258.

Barnard, J. L., 1969. The Families and Genera of Marine Gammaridean Amphipods. *Bulletin of the U. S. National Museum.* No. 271.

Brusca, G. J., 1981. Annotated Keys to the Hyperiidae (Crustacea: Amphipoda) of North American Coastal Waters. *Technical Reports of the Allan Hancock Foundation,* University of Southern California, Los Angeles. Vol. 5.

Cheng, L. and J. H. Frank, 1993. Marine Insects and Their Reproduction. *Oceanography and Marine Biology, Annual Reviews.* Vol. 31.

Cornwall, J. E., 1951. The Barnacles of California (Cirripedia). *Wasmann Journal of Biology.* Vol. 9.

Daly, K. L. and C. Holmquist, 1986. A Key to the Mysidacea and Euphausiacea of the Northeast Pacific. *Canadian Journal of Zoology.* Vol. 64.

Davis, C. C., 1949. The Pelagic Copepods of the Northeastern Pacific Ocean. *University of Washington Publications in Biology.* Vol. 14.

Dawson, J. K. and G. Knatz, 1980. Illustrated Key to the Planktonic Copepods of the San Pedro Bay, California. *Technical Reports of the Allan Hancock Foundation,* University of Southern California, Los Angeles. Vol. 2.

Laubitz, D. R., 1970. Studies on the Caprellidae (Crustacea, Amphipoda) of the American North Pacific. *Canadian Publications in Biological Oceanography.* Vol. 1.

McClain, J. C., 1968. The Caprellidae (Crustacea: Amphipoda) of Western North America. *Smithsonian Institution, U. S. Natural History Museum.* Bulletin No. 278.

Monk, C. R., 1941. Marine Harpacticoids from California. *Transaction of the American Microscopical Society.* Vol. 60.

Newman, W. A. and A. Ross, 1975. Revision of the Balanomorph barnacles; Including a Catalog of the Species. *San Diego Society of Natural History, Memoires.* Vol. 8.

Schmitt, W. L., 1921. The Marine Decapod Crustaceans of California. *University of California Publications in Zoology.* Vol. 23.

Schultz, G. A., 1969. *How to Know the Marine Isopod Crustaceans.* Wm. C. Brown, Dubuque, Iowa.

Sieg, J. and R. N. Winn, 1981. The Tanaidae (Crustacea; Tanaidaceae) of California. *Proceedings of the Biological Society of Washington.* Vol. 94.

Zimmer, C., 1936. California Crustacea of the Order Cumacea. *Proceedings of the U.S. National Museum.* Vol. 83.

Class Insecta-the insects

Cheng, L. and J. H. Frank, 1993. Marine Insects and Their Reproduction. *Oceanography and Marine Biology Annual Reviews.* Vol. 31.

Usinger, R. L., 1956. *Aquatic Insects of California.* University of California Press, Berkeley.

Class Pycnogonida-the sea spiders

Hedgpeth, J. W., 1941. A Key to the Pycnogonida of the Pacific Coast of North America. *Transactions of the San Diego Society of Natural History.* Vol. 9.

PHYLUM ECTOPROCTA-Moss Animals

Osburn, R. C., 1950-1953. Bryzoa of the Pacific Coast of America. *Allan Hancock Pacific Expeditions.* Vol. 14.

Pinter, P. 1969. Bryozoan-algal Associations in Southern California Waters. *Bulletin of the Soutnern California Academy of Sciences.* Vol. 68.

Phylum Phoronidea

Marsden, J. R., 1959, Phoronidea from the Pacific Coast of North America. *Canadian Journal of Zoology*. Vol. 37.

Phylum Echinodermata-Starfish, Sea Urchins, etc.

Boolootian, R. A. and D. Leighton, 1966. A Key to the Species of Ophiuroidea (Brittle Stars) of the Santa Monica Bay and Adjacent Areas. *Contributions to Sciences, Los Angeles County Museum*. No. 93.

Clark, H. L., 1948. A Report of the Echini of the Warmer Eastern KPacific, Based on the Collections of the Velero *III*. Allan Hancock Pacific Expeditions. Vol. 8.

Fisher, W. K., 1911-1930. Asteroidea of the North Pacific Ocean and Adjacent Waters. *Bulletin of the U. S. National Museum*. Vol. 76.

Furlong, M. and V. Pill, 1972. *Starfish*. Ellis Robinson Publishing Co.

Phylum Chordata
Class Urochordata—The Tunicates

Van Name, W. A., 1945. The North and South American Ascidians. *Bulletin of the American Museum of Natural History*. Vol. 84.

Classes Chondrichthyes and Osteichthyes-The Fishes

Cross, J. N. and L. G. Allen, 1994. Fishes. In: Dailey, M. D., D. J. Reish and J. W. Anderson, *Ecology of the Southern California Bight*. University of California Press.

Love, R. M., 1991. *Probably More Than You Want to Know About the Fishes of the Pacific Coast*. Really Big Press, Santa Barbara, CA.

Miller, D. J. and R. N. Lea, 1972. Guide to the Coastal Marine Fishes of California. *California Department of Fish and Game*, Fish Bulletin, No. 157.

Class Aves-The Birds

Baird, P. H., 1994. Birds. In: Dailey, M. D., D. J. Reish and J. W. Anderson, *Ecology of the Southern California Bight*. University of California, Berkeley.

Peterson, R. T., 1990. *A Field Guide to Western Birds*. 3rd edition. Houghton Mifflin Co., Boston.

Wilbur, S. R., 1987. *Birds of Baja California*. University of California Press.

Class Mammalia-The Mammals

Bonnell, M. J. and M. D. Dailey, 1994. Marine Mammals. In: Dailey, M. D, D. J. Reish and J. W. Anderson, editors. *Ecology of the Southern California Bight*. University of California Press, Berkeley.

Ingles, I,. G., 1965. *Mammals of the Pacific States*. Stanford University Press, Stanford.

Orr, R. T. and R. C. Heim, 1989. *Marine Mammals of California*. University of California Press, Berkeley.

PLANT KINGDOM
Algae

Abbott, I. A. and G. J. Hollenberg, 1976. *Marine Algae of California.* Stanford University Press.

Abbott, I. A. and E. Y. Dawson, 1978. *How to Know the Seaweeds.* 2nd edition. W. C. Brown Publishers.

Dawson, E. Y., 1966. *Seashore Plants of Southern California.* University of California Press.

Murray, S. N. and R. N. Bray, 1994. Benthic Macrophytes. In: Dailey, M. D., D. J. Reish and J. W. Anderson, editors. *Ecology of the Southern California Bight.* University of California Press, Berkeley.

Stewart, J. G., 1991. *Marine Algae and Seagrasses of San Diego County.* Publication of the California Sea Grant College, University of California Press.

Flowering Plants

Faber, P. M., 1982. *Common Wetland Plants of Coastal California.* Pickleweed Press.

Jepson, W. L., 1970. *A Manual of the Flowering Plants of California.* University of California Press, Berkeley.

Munz, P. A., 1974. *A Flora of Southern California.* University of California Press, Berkeley.

INDEX

215